## Lecture Notes in Geoinformation and Cartography

Publications of the International Cartographic Association (ICA)

Series Editors: William Cartwright, Georg Gartner, Liqiu Meng, Michael P. Peterson



For further volumes:
http://www.springer.com/series/7418

László Zentai • Jesús Reyes Nunez
Editors

# Maps for the Future

## Children, Education and Internet

Springer

*Editors*
László Zentai
Eötvös University
Dept. of Cartography and
Geoinformatics
Budapest
Hungary
laszlo.zentai@elte.hu

Jesús Reyes Nunez
Eötvös University
Dept. of Cartography and
Geoinformatics
Budapest
Hungary
jesus@map.elte.hu

ISSN 1863-2246 e-ISSN 1863-2351
ISBN 978-3-642-42957-6 ISBN 978-3-642-19522-8 (eBook)
DOI 10.1007/978-3-642-19522-8
Springer Heidelberg Dordrecht London New York



Printed on acid-free paper

Springer is part of Springer Science+Business Media (www.springer.com)

# Contents

# Chapter 1
# International Associations and the Provision of Outreach Programmes for Education and Training

**William Cartwright**

**Abstract** The cartographic profession has changed to one that is supported by contemporary digital production, storage and distribution devices and communication resources. What also changed are the organisations that conduct mapping programmes and 'build' repositories of geographical knowledge, digital and material. Long gone are the days of large governmental mapping agencies that had their 'tried and true' methods of data capture, processing and dissemination. Today is the workplace of small government agency; contractors (large and small); regional, national and global publishing collaborations; and map producer/user. This, in turn, has led to changes in cartographic education courses, from what could be described as focused courses to more generalist courses. Gone are the days when a graduate could accommodate the in-house practices and procedures on day 1 of a job. Everything has changed, but the underlying need for useful (and usable), accurate and timely geospatial products remains as the essential underpinnings for what we do.

In order for students to have access to relevant courses and for industry to keep abreast with developments in technology and contemporary cartography and GI Science thinking it is important for relevant educational courses to be offered. This can be done through face-to-face courses or via on-line delivery. The International Cartographic Association (ICA) is committed to supporting existing educational courses and providing specialist courses where needed. This paper provides an overview of the ICA's strategies towards the provision of education, internationally. It also gives examples about how educational courses have been presented by the ICA's international cartography and GI Science community.

W. Cartwright (✉)
School of Mathematical and Geospatial Sciences, RMIT University, Melbourne, VIC, Australia
e-mail: william.cartwright@rmit.edu.au

L. Zentai and J.R. Nunez (eds.), *Maps for the Future*,
Lecture Notes in Geoinformation and Cartography 5,
DOI 10.1007/978-3-642-19522-8_1, 

## 1.1 Introduction

The International Cartographic Association is the world authoritative body for cartography, the discipline dealing with the conception, production, dissemination and study of maps. The ICA was founded on June 9, 1959, in Bern, Switzerland. The first General Assembly was held in Paris in 1961. The mission of the International Cartographic Association is to promote the discipline and profession of Cartography and GIScience in an international context.

The activities of the ICA are important for promoting and advancing the theory and praxis of cartography. Throughout its 50-year history, ICA has brought together researchers, government mapping agencies, commercial cartographic publishers, software developers, educators, earth and environmental scientists, and those with a passion for maps.

The International Cartographic Association exists:

- To contribute to the understanding and solution of world-wide problems through the use of cartography in decision-making processes.
- To foster the international dissemination of environmental, economic, social and spatial information through mapping.
- To provide a global forum for discussion of the role and status of cartography.
- To facilitate the transfer of new cartographic technology and knowledge between nations, especially to the developing nations.
- To carry out or to promote multi-national cartographic research in order to solve scientific and applied problems.
- To enhance cartographic education in the broadest sense through publications, seminars and conferences.
- To promote the use of professional and technical standards in cartography.

The Association works with national and international governmental and commercial bodies and with other international scientific societies to achieve these aims.

*(Adopted by the 10th General Assembly of the International Cartographic Association, Barcelona, Spain, 3 September 1995.)*

An important contribution that ICA makes through its international community is outreach and technology transfer. This is supported through direct ICA initiatives, the activities of Commissions and Working Groups and programmes conducted with ICA Affiliates.

## 1.2 The ICA Strategic Plan

The initial ICA Strategic Plan 2003–2011 (ICA 2003) provided a number of guidelines for implementing plans for both Education and Professional Practice. In July 2011, at the ICA General Assembly in Paris, a new Strategic Plan for

2011–2019 was approved. In Appendix 3 of this new Strategic Plan, entitled "Using the Strategic Plan to create Operational Plans", several objectives for ICA related to professional practice and education were proposed. (These objectives complemented other objectives related to Science, Society and the Arts) (ICA 2011).

### *1.2.1 Professional Practice*

The Strategic Plan for 2003–2011 included that statement of: "Amateur and professional practice within the Geospatial sciences will change in nature, increasing the necessity for Continuing Professional Development". In Appendix 3 of the ICA Strategic Plan for 2011–2019 several objectives for ICA related to professional practice were proposed, viz:

- To encourage wider application of Cartographic principles within information technology.
- To promote the transfer of GI technology and standards for professional use.
- To strengthen the profile of professional practice commissions in ICA.
- To promote the presentation of 'best practice' in the field of Cartography and GI science.
- To provide possibilities for interaction between practitioners during the ICA conferences.

This was to be formalized through several actions:

- Analyse commission structure and propose new commissions in order to maintain a balance between theory and practice.
- Organise workshops on specific topics.
- Organise high quality technical exhibitions and expert panels during the conferences to attract practitioners to participate and exchange ideas.
- Encourage national associations and universities to translate proceedings of conferences and symposia into local languages, publish the translations on the web and link them to the ICA page.
- Facilitate the exchange of experts between and within developed and developing countries and revive the ICA 'Third World' policy.
- Support United Nations by providing geographic expertise.

### *1.2.2 Education*

From Appendix 3 of the ICA 2011–2019 Strategic Plan, it was proposed that the ICA should:

- Investigate ways to strengthen and monitor education programmes in Cartography, GI.

- Science and related subjects at all levels (university, high school, elementary, life-long learning).
- Investigate fora for discussions of education programs and curricula in Cartography and
- GI science.
- Develop information networks and online courses on Cartography and GI science.
- Offer educational courses for students on Cartography and GI science for example in
- Developing countries and for regional purposes.
- Offer 'master classes' in GI Systems/mapping to guide managers in spatial decision making.
- Investigate methods (and funding sources) to encourage the participation of students and other young members in ICA activities.

It proposed the following actions:

- Analyse existing university curricula in Cartography and GI science.
- Help widen the Cartographic/GI science knowledge base and skills into new segments of Society.
- Increase efforts directed to capacity-building, especially in developing countries, especially with reference to human resource development.
- In co-operation with commercial suppliers, develop online courses on Cartography and GI science to support and complement existing courses.
- Facilitate provision of geographic data for educational use.
- Support appropriate United Nations activities by providing geographic expertise.

This chapter reports on some of the actions being carried out by the ICA, its member nations, Commissions and Working Groups and affiliates to advance these objectives.

## 1.3 The International Cartographic Association and Education

From its very beginnings, the International cartographic Association was committed to provide training and education in cartography. At the 1964 General Assembly the Association selected three recommendations for Commissions for further development: Training, Terminology and Automation (Ormeling 1987, 1988).

The ICA's Strategic Plan proposes a number of actions related to Ideas and Actions for the Organisation and in the Wider Operational Environments. Some of these actions that relate to education are:

- To help widen the Cartographic/GIScience knowledge base and skills into new segments of society

- To increase efforts directed to capacity-building, especially in developing countries, especially with reference to human resource development
- In co-operation with commercial suppliers, to develop virtual academy courses on Cartography and GIScience to support and complement what is on offer.

The ICA addresses these and other issues through direct ICA initiatives through its Executive and member organisations and with partners from ICA affiliates, sister societies and industry. The activities of Commissions and Working Groups and programmes provide the 'powerhouse' that supports these endeavors.

## 1.4 Commissions and Working Groups Outreach Activities

To achieve its aims the ICA operates through a number of Commissions and Working Groups. Commissions and Working Groups carry out the general operations of the ICA. They address the full range of scientific, technical and social research that is the mark of ICA activity. They achieve the transfer of knowledge about Cartography and GIScience and GI Science by publishing books and special editions of journals and running workshops and educational courses. Colleagues from the ICA community conduct these workshops on a volunteer basis, generally with the support of the national member organisation of ICA or the national mapping body.

### *1.4.1 Courses and Workshops by ICA Commissions and Working Groups*

ICA Commissions and Working Groups have conducted many outreach courses. Here, examples of some of the courses are provided.

The ICA Commission on Education and Training has been at the forefront in the provision of training and education courses in Cartography. The first Commission Chair was *Stephane de Brommer*, from IGN, France. He chaired the Commission from 1964 to 1972 (ICA CET 2011). His successors have continued the tradition of providing quality courses, face-to-face and via publications and contemporary communications systems like the Internet.

Recent initiatives of the ICA Commission on Education and Training have included in a workshop on Cartography in Tehran, Iran in May 2009. This was undertaken in collaboration with the National Cartographic Center of Iran, ran. Figure 1.1 shows some of the 30 participants at the workshop.

This course followed a successful course run in 2008—hands-on web-mapping technologies—conducted by the ICA Commission on Maps and the Internet and organised by the National Cartographic Center of Iran. The workshop focused on the technological and methodological basics of delivering maps on the web,

**Fig. 1.1** Participants in the workshop on cartography, Tehran, Iran, May 2009 (Photograph courtesy ICA Commission on Education and Training)

including such topics as basic tools, design questions, interactive functions and using map server technology.

In July/August 2010 the ICA Commission on Education and Training ran training courses in Ho Chi Minh, Vietnam and Jakarta, Indonesia. The workshops were conducted to facilitate technology transfer. The Ho Chi Minh City Workshop, 19–23 July 2010, covered Cartographic Information Systems (GIS with a cartographic emphasis), Internet Mapping, Atlas production, Prepress and Digital Printing. The second week of the course, which addressed Web Cartography and its applications. It covered the topics of Web mapping, Web 2.0 and the design and production of map mash-ups (Fig. 1.2).

The second workshops were held in Jakarta, Indonesia. In the first week the topics covered were Map projections, Map design/Cartography design, Cartography theory (Map Evaluation), Atlas cartography, Toponymy, Map production using GIS, Prepress design and layout using GIS, Computer assisted cartography using ArcGIS and 3D GIS. In week 2 a program on Web cartography was conducted (Fig. 1.3).

The ICA Commission on Management and Economics of Map Production has regularly organised workshops in Urumchi and at the Intercarto Conference and an ICA-sponsored workshop in Gent in 2009.

Through their input in the United Nations Group of Experts on Geographical Names, ICA cartographers have regularly participated in the toponymy course programme of UNGEGN. Courses have been held in Khartoum (2003), Bathurst (2004), Maputo (2004 and 2006), Malang (2005), Tunis (2007), Ouagadougou (2008), Vienna (2008), Timişoara (2008), and Nairobi (2009). The major item in these courses is the conveyance of the awareness of the importance of

**Fig. 1.2** ICA Commission on Education and Training, in collaboration with the Ho Chi Minh University of Technology and the Vietnamese society for cartography, geodesy and remote sensing. Workshop on web 2.0 and cartography—Ho Chi Minh City, 2010 (Photograph: William Cartwright)

**Fig. 1.3** Manuela Schmidt (Austria) conducting hands-on practical components of the ICA Commission on Education and Training, in collaboration with Bakosurtonal. Workshop on web 2.0 and cartography—Jakarta, 2010 (Photograph: William Cartwright)

geographical names as part of a nation's spatial data infrastructure, and the need to collect these names correctly and efficiently for use on maps and in gazetteers (Fig. 1.4).

The ICA Commission on Education and Training has developed a virtual course on Cartography and GIScience in collaboration with contributions from universities and individual academics. Academics and practitioners have provided the courses from the international cartographic community. They can be accessed and used free

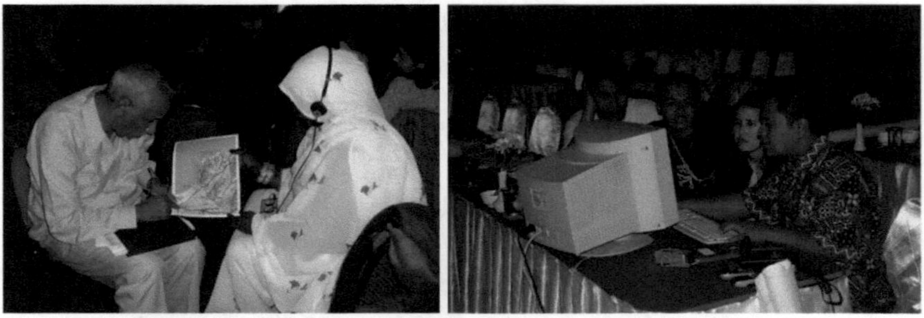

**Fig. 1.4** Participants preparing for the names-collecting fieldwork UNGEGN toponymy course. At left—Tunis 2007, at right—Malang 2005 (Photographs courtesy of Ferjan Ormeling)

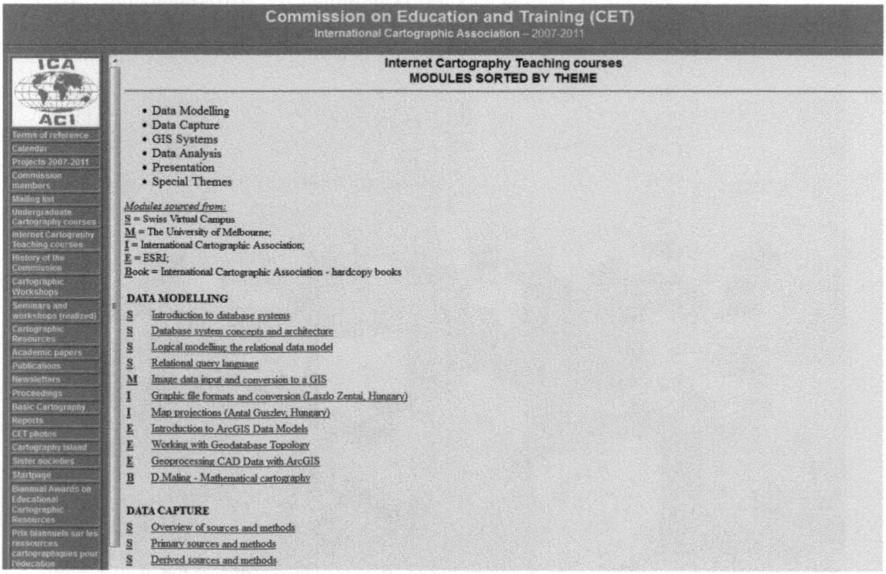


**Fig. 1.5** Commission on Education and Training on-line courses (Source: http://lazarus.elte.hu/cet/)

of charge. The courses can be accessed via the Commission Web site at: http://lazarus.elte.hu/cet/. The image in Fig. 1.5 shows the interface to the courses that are offered on-line.

The ICA Working Group on Open Source Geospatial Technologies promotes multi-national holistic research in free and open source geospatial technologies in order to make accessible the latest developments in open source tools to the wider cartographic community. The WG attempts to enhance the usage of free and open source geospatial tools among the cartographic community worldwide, especially for education. The WG organises workshops with the aim to capacity building

**Fig. 1.6** ICA working group on open source geospatial technologies course on open source geospatial software (Photograph courtesy of Suchith Anand)

participants by providing hands on experience to develop skills in the application of open source geospatial software (Fig. 1.6).

### *1.4.2 Courses in Collaboration with Industry*

In collaboration with ESRI, Inc., a major sponsor of ICC2009 in Santiago-Chile, the Cartography with *ArcGIS* course was taught after the ICC2009 in November 2009. Mr. Makram Murad-al-shaikh, a Senior Instructor in GIS and Cartography at ESRI's Educational Services Department, taught the course. The course was offered free to all candidates attending the ICC2009 conference and later was opened to lead Chilean cartographic organisations. Nineteen attendees were trained for 3 days on both basic cartographic design principles together with hands-on training on ESRI's *ArcGIS* software tools for mapping design and production. All of the cartographic tools available in ArcGIS were explored with best practices taught on how to use them in hands-on exercises (Fig. 1.7).

### *1.4.3 Seminars for Students*

In November 2009, in collaboration with SNIT (Sistema Nacional de Coordinación de Información Territorial), Chile, members of the ICA Commission on Geospatial

**Fig. 1.7** ESRI's *ArcGIS* course, Santiago, Chile, November 2009 Photograph courtesy of Makram Murad-al-shaikh)

**Fig. 1.8** ICA Commission on Geospatial Data Standards presented a seminar for Cartography and GIScience students, Santiago, Chile, November 2010 (Photograph courtesy of Antony Cooper)

Data Standards presented a seminar for Cartography and GIScience students and staff at Universidad Technológica Metropolitana del Estado de Chile (UTEM) in Santiago, Chile. Presentations covered the areas of 'INSPIRE', quality standards for the spatial data modeling, 'SNIT' national spatial data infrastructure of Chile and standards for geographical information. Participants at the seminar are shown in Fig. 1.8.

### *1.4.4 Graduate Students' Seminars*

At the first ICA Symposium on Cartography for Central and Eastern Europe, in Vienna, Austria, in February 2009 included a special session where junior scientists

**Fig. 1.9** Panel session at the Ph.D./Master Forum at the CEE Symposium 2009, Vienna. From left Prof. Dr. Georg Gartner, Prof. Dr. William Cartwright, Prof. Dr. Necla Ulugtuken, Prof. Dr. Ferjan Ormeling and Prof. Dr. Milan Konečný (Photograph courtesy Géza Simon)

presented their research work in a dedicated Ph.D./Master Forum. Ph.D. students presented the results of their research and a panel of experienced researchers provided feedback to student presenters (Fig. 1.9).

## 1.5 Education for Children

The Barbara Petchenik Children's World Map Competition is organised every second year to coincide with biennial International Conference of the Association. This is a map design competition for children ages 15 years and younger. It is held to honor of the late Dr. Barbara Bartz Petchenik, a past Vice President of the ICA who was extremely interested in maps for children and children as cartographers. The competition is organised by the ICA Commission on Children in Cartography.

This competition has been taken-up by teachers around the world to involve their students in the world of mapping. One example of the local support to children to enter the competition in South Africa, where staff members from the Department of Geography of UNISA (University of South Africa) in Pretoria visited schools throughout the country to liaise with teachers and children and explain the competition theme—"Living in a globalized world". The photograph in Fig. 1.10 shows Professor Elri Liebenberg, Chair of the ICA Commission on the History of Cartography, and former ICA Vice-President, in one of these classes conducted by UNISA.

The ICA Commission on Children in Cartography has worked with ESRI Press to publish books containing winning entries from the competition. Two books have been published. *Children Map the World: Selections from the Barbara Petchenik Children's World Map Competition* (Jacqueline Anderson, Jeet Atwal, Patrick Wiegand and Alberta Auringer Wood editors) was published in 2005. And, in November 2009 a second book—*Children Map the World* (Temenoujka Bandrova, Jesus Reyes Nunez, Milan Konečný and Jeet Awal eds.) followed. The books

**Fig. 1.10** Mapping class for children at the Department of Geography at the University of South Africa (Photograph courtesy of Elri Liebenberg)

feature selected drawings from the Barbara Petchenik Children's World Map Competition.

## 1.6 Conclusion

The activities of the ICA are important for promoting and advancing the theory and praxis of cartography. Throughout its 50-year history, ICA has brought together researchers, government mapping agencies, commercial cartographic publishers, software developers, educators, earth and environmental scientists, and those with a passion for maps. The Cartography and GIScience world has changed significantly since 1959—the role and impact of ICA has been steadfast. Its mission is to support and promote Cartography and GIScience– globally. Its outreach programmes, in many instances conducted with national member organisations, affiliates and industry, are conducted to contribute to the transfer of knowledge and to foster the advancement of the discipline.

**Acknowledgements** This chapter was prepared with inputs from colleagues from the ICA international community. Thanks to Dr. Suchith Anand (Chair, ICA Open Source Geospatial Technologies Working Group), Dr. Antony Cooper (Chair, ICA Geospatial Data Standards Commission), Professor Dr. Philippe de Maeyer (Chair, ICA Management and Economics of Map Production Commission), Dr. David Fairbairn (former ICA Secretary-General/Treasurer), Assoc. Professor Dr. David Fraser (Chair, ICA Education and Training Commission), Professor Dr. Georg Gartner (President ICA), Professor Elri Liebenberg (Chair, ICA History of Cartography Commission and former Vice-President of ICA), Makram Murad-al-shaikh (ESRI, Inc.), Professor Dr. Ferjan Ormeling (former Secretary-General/Treasurer ICA, Vice-Chairman of the UN Group of Experts on Geographical Names (UNGEN) and Convenor of the UNGEGN Working Group for Training Courses in Toponymy) and Professor Dr. Michael Peterson (former Chair, ICA maps and the Internet Commission).

## References

ICA (2003) Strategic plan http://www.icaci.org/documents/reference_docs/ICA_Strategic_Plan_2003-08-16.pdf

ICA (2011) Strategic Plan, 2011–2019 http://www.icaci.org/documents/reference_docs/ICA_Strategic_Plan_ 2011-2019.pdf

ICA Commission on Education and Training (2011) History of the CET. http://lazarus.elte.hu/cet/

Ormeling FJ (1987) ICA 1957–1984. The first twenty-five years of the International Cartographic Association. Enschede, ICA 1987

Ormeling FJ (1988) ICA 1959–1984. The first twenty-five years of the International Cartographic Association. Elsevier, Enschede, ICA 1988

# Chapter 2
# Exercises in Cartography

**Ferjan J. Ormeling**

**Abstract** By describing the scope and intent of contemporary cartographic exercises, the author pictures the changes that took place in cartography during the last 50 years in the format of an autobiography. Although these exercises changed from manual and repetitive ones towards digital and unique tasks, during this development some freedom of expression was lost, as current GIS-oriented software packages limit the design options. The author calls for exercises where the geographical insight of cartography students is strengthened, where they are made aware of the bandwidths of cartographic license and of the existence of many different but valid ways of rendering the same realities. Also, through exercises where they are addressed as map users, trainees may realise the map-use impact of the graphical decisions taken and the real challenges cartographers face in visualizing geospatial data for decision support.

## 2.1 Cartographic Apprentice

When I first started as an apprentice atlas editor, in 1961, I had to learn how to apply lettering to maps, and I spent many evenings just drawing the letter o. After a week I graduated to variations of o, such as a, b, d, g, q, c and e or p. Then I moved on to n, and varied it with h, u and m; the next month would be focused on i and l, f, j, t and odd letters like k, r, s, v, w, y and z. The next step would be to combine these letters and to get used to the differing distances between them. I did not feel particularly enriched by these long evenings and I merely wondered whether I had opted for the right profession. Fortunately, nowadays this lettering is done digitally and cartography students won't loose time in doing lettering exercises—at least not on lettering itself, but they still have to do exercises in the application of geographical names to maps.

F.J. Ormeling (✉)
Utrecht University, Utrecht, The Netherlands
e-mail: f.ormeling@uu.nl

L. Zentai and J.R. Nunez (eds.), *Maps for the Future*,
Lecture Notes in Geoinformation and Cartography 5,
DOI 10.1007/978-3-642-19522-8_2, © Springer-Verlag Berlin Heidelberg 2012

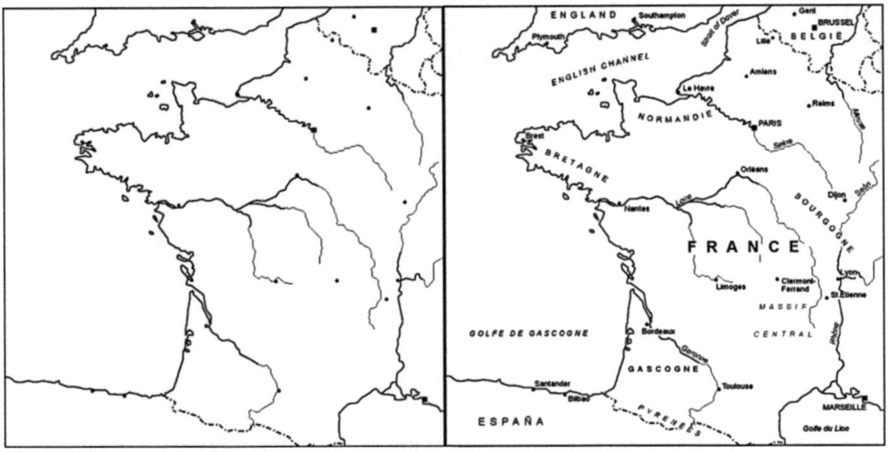


**Fig. 2.1** Exercise in map lettering from the Basic Cartography Exercise Manual (1991)

Here they have to apply the theories of Imhof, Bonacker or Spiess to the map, in order to make sure that there is no ambiguity regarding the symbols a name refers to, to ensure the shortest possible time to find a name on the map, by using variations in type styles, sizes, boldness, spacing and colour (Fig. 2.1).

In my cartography classes at the university at the same time I had to be able to operate geodetical or photogrammetrical equipment, and map contour lines from pairs of aerial photographs, using both hands and feet. So I had to do mapping exercises with the equipment. Again, this was not particularly stimulating. As my main subject was geography we also had to engage in statistical exercises, do sums in order to compute the number of observations needed to end up with results that were 99% significant, or compute different kind of averages. The course in statistics was used as a threshold to keep out those without a head for mathematics although I never later on perceived any need why such a head was required in geography, nor in cartography. Fortunately for me, the looming onset of computers had obliterated—in the mind of the teaching staff—the need to do exercises in the plotting of map projections, with coordinatographs and the assistance of goniometric tables. As part of geography we also had to do geomorphology courses, and here we had to learn to draw all kinds of diagrams, cross sections, block diagrams and longitudinal sections. As we were interested in visualization we taught ourselves how to draw panoramic maps, based on topographic maps (Fig. 2.2).

## 2.2 Cartography Courses

But the real interest in cartographic exercises only started when I had graduated and, as a member of Utrecht university staff, helped to start up a Master's programme in cartography in 1971. Then we got to know a French publication called *La cartographie*

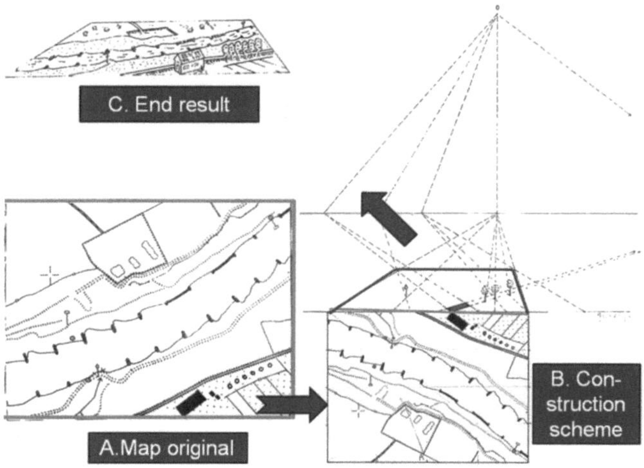


**Fig. 2.2** Producing a panorama from a topographic map

*thématique comme méthode de recherche*, by Claval and Wieber (1969), which finally presented some intellectual challenge. It contained statistics on point, linear or areal data and base maps on which these data had to be visualized. Here at least there also was an opportunity to apply one's geographical knowledge, as doing these exercises frequently called for judgment regarding what was most important from a geographical point of view. As an example, in Fig. 2.3, I show an assignment to map statistical data on agriculture for Indonesia, where one has to pick the most important aspect of the table (based on one's geographical knowledge of the country) and visualize it.

The textbook by Claval and Wieber also showed us that—at least within Europe—there were different schools of cartographic design, using different mapping techniques. For instance, in France in the 1970s the areal cartogram method devised by Aimé Perpillou was en vogue. This was a quite elaborate method of showing different percentages of land use, per enumeration unit, by using coloured bands that together made up 100% (Fig. 2.4). The method called for extreme generalization of the data and the resulting images were not altogether straightforward. This method was never used in Germany or in the United Kingdom. In Germany, isoline maps were only used for physical phenomena, and the United Kingdom was an early advocate of anamorphosis maps, which were rather frowned upon in Germany. And it was only in Czechoslovakia that a particular time-related diagram type ever occurred.

## 2.3 ICA's Basic Cartography Programme

In the 1970s, there was already an ICA project under way to produce a textbook for cartographers. This endeavour was supported by UNESCO, and it resulted in the *Basic Cartography* manual series (Anson and Ormeling, 1984-2002), containing three

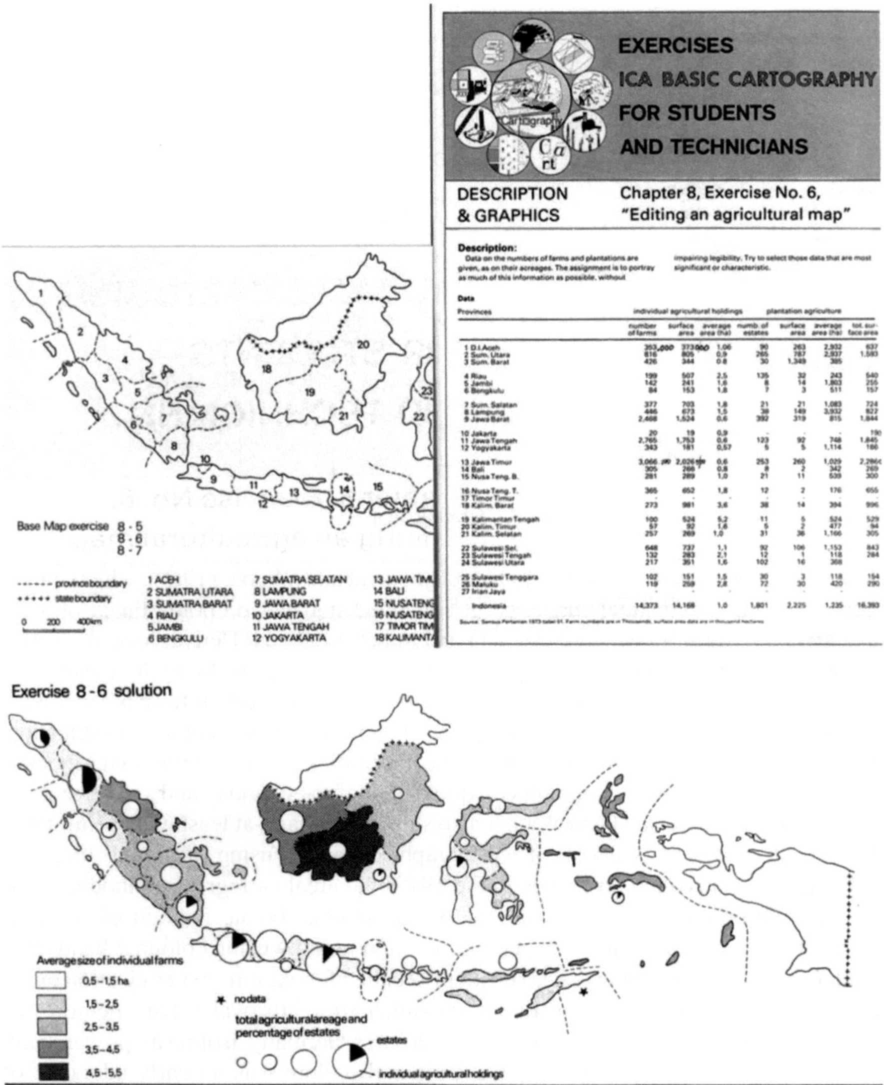


**Fig. 2.3** Exercise on agricultural mapping with statistical data and base map (upper part). For solution see lower part. From the *Basic Cartography Exercise manual* (1991)

manuals (with contributions from France, FRG, GDR, Israel, Japan, the Netherlands, Nigeria, UK, USA and Sweden) and an exercise book. I travelled all over Europe to visit cartographic establishments in order to solicit 500 printed copies each of their best exercises, framed in a uniform ICA template and to put together this exercise manual, which finally consisted of contributions from, Austria, Belgium, Canada, France, Germany, Hungary, Israel, Japan, the Netherlands, Sweden and Switzerland.

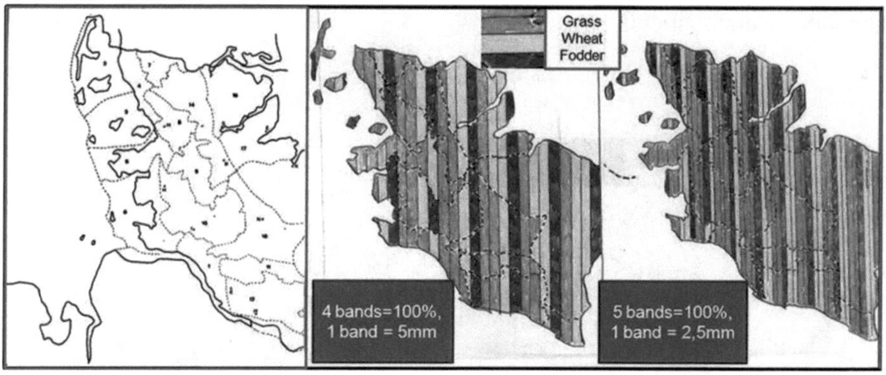


**Fig. 2.4** Diagram map following the Aimé Perpillou method (From Claval and Wieber (1969))

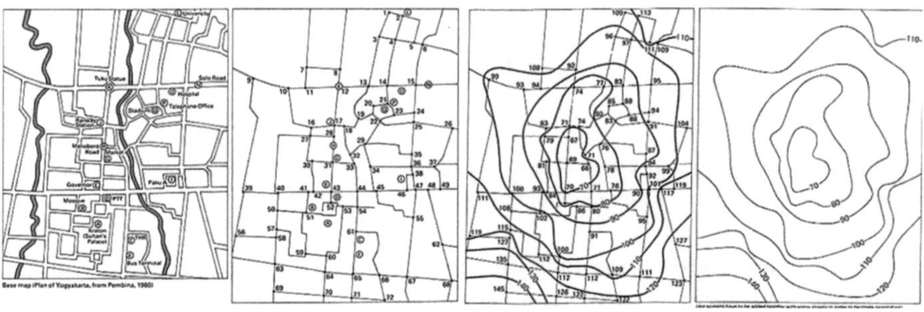

**Fig. 2.5** Map exercise to assess the accessibility of urban services; the cumulative number of arcs it takes to get from each corner point to all public facilities is used as input for an accessibility isoline map (From Basic Cartography Exercise Manual (1991))

Apart from map construction exercises, the geographers in our faculty also had to do map use exercises, for analytical purposes, and these existed for instance of assessing accessibility, nearest neighbour values or quantifying patterns and shapes (see Fig. 2.5, for instance).

What is especially significant in doing most of these exercises is that the students see the effects of their design decisions by comparing their results with those of their colleagues and perceive how these differences affect or boost the information transfer: which method is best, pie graphs or columns to render absolute statistical values? Is it better to render the absolute values of the data or to link them to other data, so as to normalize them? (See Fig. 2.6). In a dot map exercise, where a base map and a description of the residential housing characteristics for a campus would be provided and the students asked to work out the best representative value and size of the dots with which to render the population distribution pattern, again comparing the results always was as much of an eye-opener to the students as doing the exercise themselves.

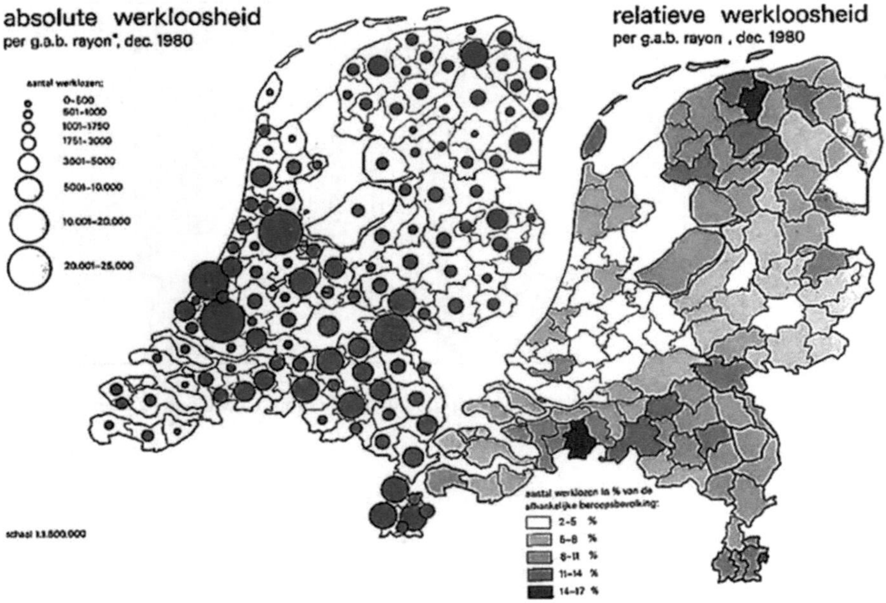


**Fig. 2.6** Unemployment in the Netherlands, left rendered with proportional symbols, and at right normalized, i.e. expressed as a proportion of the total labour force

Next to map lettering, the most important exercises before the advent of the computer surely were those of rendering information according to a standardised legend, in which all the symbols and their measurements, and the various line widths were prescribed in a master legend. This part was rather repetitive, tiresome and disappointing, as the teachers always seemed to be able to discern who had drawn which map, in spite of us all adhering to the prescribed legend and specifications.

As my country is rather small there seemed not to be enough potential cartographic draughtsmen to start a regular cartography course for, so we opted for a correspondence course, in which the participants had to do exercises and had to send them in, and would receive them back within a week, annotated with the comments of the teachers (PBNA cursus, 1973-1976). On the exercise sheets the planimetry to be inked in was rendered in blue, and with special drawing pens and India ink the map detail had to be applied. In my first presentation to an ICA conference, in Moscow, in 1976, I reported on our experience with setting up this correspondence course.

## 2.4 Meta-exercises

Next to map drawing and map lettering came map generalization, and here it was again important that one developed a feeling of hierarchy, a feeling of what was important and should be retained, even if this activity was closely outlined as well,

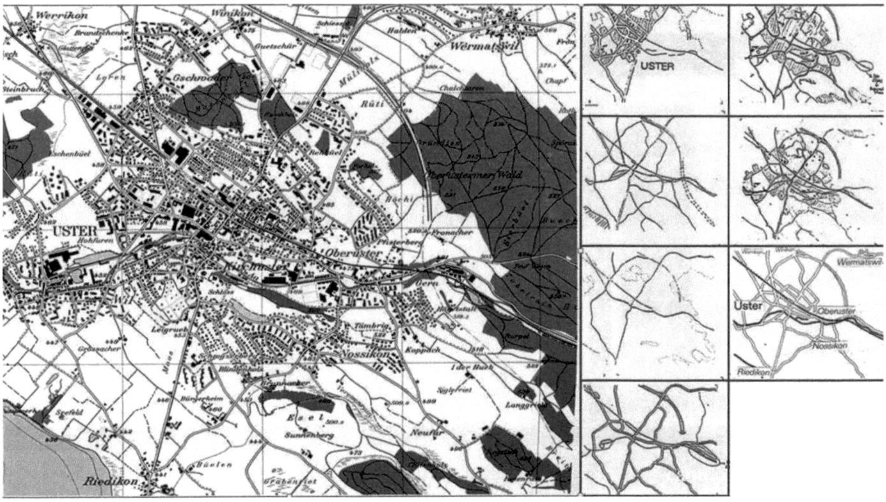


**Fig. 2.7** Generalisation exercise (by Prof. E. Spiess, *Basic Cartography Exercise manual*, 1991) with different results

so as to render the differences between the results of individual cartographers as small as possible. Legend-sheets with the exact dimensions of the symbols and lines to be used served as examples but, even when these specifications were adhered to as strictly as possible, students still ended up with a wide range of different results. When we confronted the students with the variation in their results, there was a meta-effect in this exercise—as the students could see that the results of their colleagues would deviate to some degree from their own, and this showed them a bandwidth of cartographic licence, it showed them that different results based on the same original data still were acceptable to a certain degree (Fig. 2.7).

Another drawing technique that had to be mastered through exercises where some geographic knowledge came in handy was hill-shading, which we would do with pencil, but which Swiss and Austrian experts would apply with airbrushes.

Picking the correct colour schemes or colour ramps is another item to do exercises in, in order to make sure that students see for themselves which colour combinations work or not, when diverging hues should be opted for, or how a colour ramp can be extended to accommodate a larger number of classes. Cynthia Brewer here helps students doing exercises on these items enormously with her Colorbrewer website (http://colorbrewer2.org).

For myself I have found much profit for the students in an exercise in which they are asked to render a same phenomenon, using the same figures in as threatening a way and in as inconspicuous a way, just by manipulating colours and class boundaries. By doing this exercise they realize the impact of their selection of class boundaries and colour schemes.

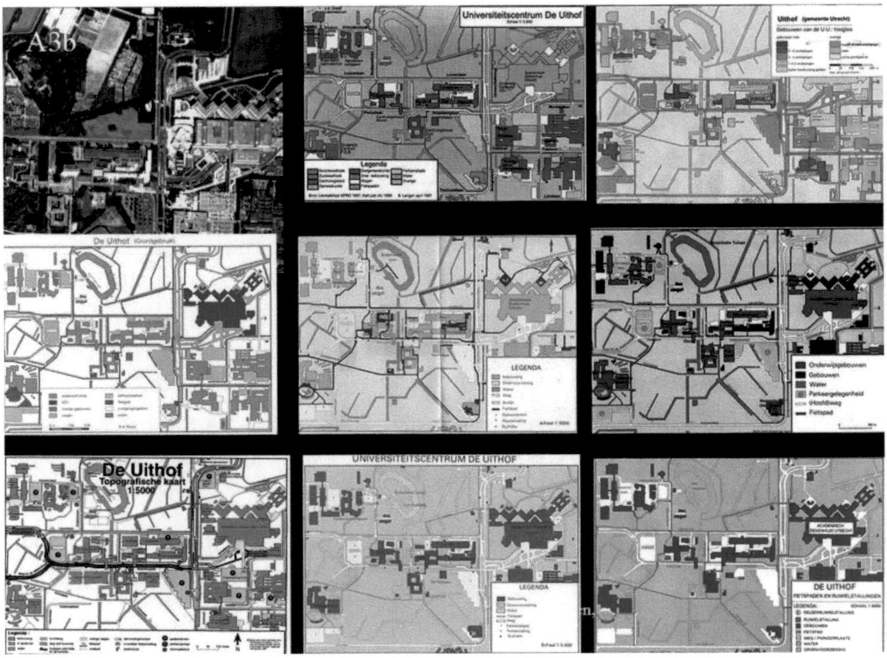


**Fig. 2.8** Student maps of Utrecht University campus, based on aerial photograph (top left)

For the map reproduction classes we had in Utrecht University in the 1980s, prior to the advent of digital techniques, we did a simple exercise to produce a map of the university campus. Students could select their own target audience for this campus map, and customize the spatial information to convey accordingly. Here again the results are most instructive when compared. That is because they show how—even with very simple data, as on the function of buildings, their number of floors or the transport facilities—a host of different images will emerge, based on different decisions regarding information hierarchy, preferential colour schemes or student outlook—even differences between car-owning students and public transport advocating students would stand out on the maps (Fig. 2.8).

Part of the Utrecht cartography programme consisted of fieldwork, in France, where one of the assignments was to update an existing topomap, and another to do a land use survey of a small area. Apart from the generalization aspect, the results here provided another example of the impact of individual land use classification decisions, and of the generalization rules followed, even if the assignment—to produce a land-use based map 1:15.000 for the area for cyclists and hikers—would let us to expect rather homogeneous results. The realization that the same reality, to be mapped to the same specifications, could be reflected in so many different, but valid views of the same reality certainly was an eye-opener for the students.

## 2.5 Computer Classes

The map drawing exercises that replaced the manual ones took place on the computer, using specific drawing programmes like Aldus Freehand in the 1990s, and now perhaps Adobe Illustrator. I suppose every university cartography department would train its students with exercises to become familiar with these mapping packages, learn to deal with the various layers, at least before the advent of GIS software programmes, when boundary files could just be matched with statistical data sets, and many students do not get beyond learning how to combine data sets in EXCEL and deal with shape files in ArcGIS.

Of course the new digital environment also affected reproduction. The reproduction exercises contained in the ICA Basic Cartography Exercise manual, like those on the production of small-scale topographic maps, or of tourist maps no longer were relevant, and would have to be restructured. The same was valid for the production diagrams. A fair amount of time used to be spent on exercises on their compilation; I remember ICA workshops all over Asia where we trained students to find the least expensive ways of reproducing maps through these exercises that now had to be reworked, necessitating new sets of symbols for the new digital techniques required.

## 2.6 Atlas Production Exercises

Atlas production exercises simulate many aspects of the cartographic profession, as they would train both the design and the production planning aspects. I reported on them at the Tokyo ICA conference (1980). In Atlas production classes e.g. lay-out exercises were done, to establish a template for the individual atlas sheets, and for finding the best sequence of map subjects. An example here is the exercise how to structure a school atlas of Turkey, to be printed on both sides of a single printing sheet, one side in colour and the other in black and white. The sequence had to make sense and be thought logical; the most important maps would have to be coloured, but also the chorochromatic maps that would not be legible in black and white. For all individual atlas maps, preliminary drawings or mock-ups would be made, before starting with their digital production, in order to be sure all necessary elements would be incorporated (see Fig. 2.9).

Superior examples of atlas production exercises would be the atlases compiled at Oxford Brookes University where Roger Anson would take his students to the continent each year in order to gather the information to visualize in the next term.

Toponymy is another area of atlas cartography where exercises are used in order to speed up the student's grasp of the subject matter. We would ask them, for instance, to do an exercise in script conversion, or to compare geographical names on a map of Spain from a Spanish school atlas and from a British school atlas, and then ask them to work out the rules the editor of the British atlas would have followed in order to adjust the toponyms to his British audience. Production of a place names index and from a map would teach students how to deal with generic

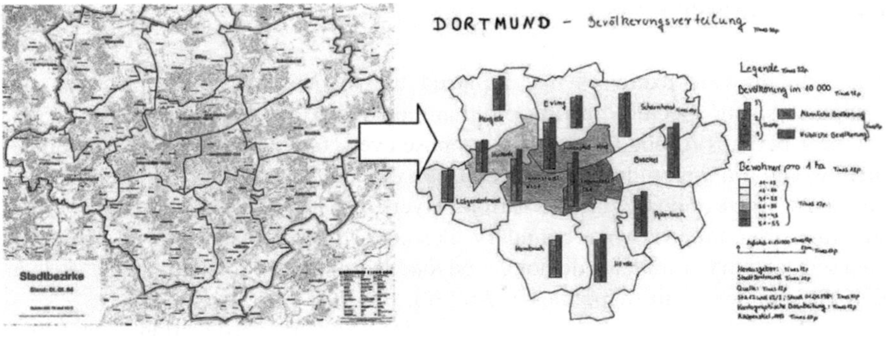


**Fig. 2.9** Mock-up for an atlas map exercise, from 'Ausbildungsleitfaden' (2000)

name parts like Rio or Cape in a names index, and what to do with homonyms and allonyms. First they would have to copy all names from a map, identifying their object category and location, and then reordering them according to a specific alphabet.

A final aspect of map production would be an annotation exercise: how to prepare one's map or atlas in such a way—through adding marginal information—that an independent librarian would be able to list all the relevant information (impressum, publisher, author, date and place of publication) for a web bibliography or catalogue.

## 2.7 Map Use Exercises

Interpretation of maps or doing measurements on them is the other half of cartography, and although map use exercises are not primarily oriented at cartographers, they are still very suitable to increase the cartographers' awareness of how map users react to their products. Doing these kinds of exercises, they would for instance find out to which degree different measurement outcomes thought to be relevant by map users are just a consequence of generalisation.

Some examples of map use exercises are e.g. how to prepare a slope map (where, per grid cell the difference between the highest and lowest point is assessed first and then converted in a slope value, assigned to the grid cell centre, and followed by interpolation to produce an isoline slope map (Warn 1980)), drainage density assessment (here the length of rivers per $km^2$ grid cell would be computed, and the grid cell data categorized) or an exercise to interpret landforms from a contour line map (Schulz 1995).

## 2.8 Map Exercises for the Future

The framework of all these cartographic exercises described here has been different in scope—there were those meant for in-house training for apprentices, there were exercises that were part of the training of geographers and exercises meant for cartographic draughtsmen. There were higher education or university cartography courses that had exercises in a classroom environment, and there were exercises that formed part of correspondence courses, or of manuals developed by cartographic societies or individual teachers. The Netherlands Cartographic Society used to invite foreign experts for its courses on specific techniques like map lettering or hill shading, for the benefit of their members who had to do exercises in order to master these techniques.

All of these exercises were meant to train, for the participants to practice the theoretical aspects, to pass on knowledge and experience. The ICA tries to continue this training work globally with the hands-on courses it has been organizing for some 30 years, not only through its Commission on Education and Training (CET) but also through the Commission on Management and Economics of Map Production and the Commission on Maps and the Internet, frequently also jointly.

As only a small audience can be reached, however enthusiast these commissions are, with lecturing teams only within the last 2 years (2009 and 2010) flying in into Central Asia, Iran, Indonesia or Vietnam, developing web courses are a way to reach larger audiences, even if interaction and feed-back still present problems. Already an enormous variety of cartography courses is being offered on the web, and the ICA Commission on Education and Training under Laszlo Zentai and David Fraser has selected the best ones and incorporated them on the CET website, a major job that deserves acknowledgement. The only aspect perhaps lacking in these web courses are exercises. I think the best and most challenging exercises that we have devised for our own students should be incorporated onto the CET website next to the current lessons or lectures that have already been stored there, just as we collected the best paper exercises in the past for the ICA Basic Cartography Exercise manual.

These best and most challenging exercises should show students how many-sided and intellectually stimulating cartography is, and they might thus be induced to follow a career in our discipline. With these exercises we would visualise the challenges of our profession, to adapt geospatial data to the objectives and the target groups of information transfer, to support spatial decision making now and in the future.

## Annex 1: Overview of Cartographic Exercises

A. Map production exercises
- A1. Information analysis
  - (1a) Establish information/parameter hierarchy
  - (1b) Establish parameter characteristics
  - (1c) establish rules for language-dependent toponymy
- A2.Exercises in mapping technique
  - (2a) Map lettering
  - (2b) Hill shading
  - (2c) Generalisation
  - (2d) Drawing panoramic maps/block diagrams
  - (2e) Line drawing exercises

(continued)

A3. Exercises in design
(3a) Selecting map types
(3b) Classification/characterization
(3c) Classification/manipulation
(3d) Colour ramp/colour scheme exercises
B. Exercises in map reproduction
B1. Devise optimal reproduction method
B2. Construct reproduction diagrams
B3. Devise optimal atlas structure (Turkey)
B4. Prepare (atlas) map annotation to enable proper documentation
B5. Prepare map names index
B6. Produce standard lay-out and specifications
B7. Produce a mock-up of an atlas (sheet)
C. Map use/analysis exercises
C1. Recognise symbology
C2. Assess accessibility
C3. Determine patterns: nearest neighbour
C4. Recognise terrain forms (geomorphology)
C5. Working with grids
C6. Find position, in degrees, minutes and seconds or in decimal degrees
C7. Convert decimal degrees into degrees, minutes and seconds
C8. Establish directions, describe way points
C9. Assess network densities
C10. Assess terrain changes in-between successive map editions
C11. Assess slopes/gradients; describe expected relief on a route
C12. Establish distances and time needed to reach a destination
C13. Compute areas and distances
C14. Establish profiles, visibility analyses
C15. Establish spatial association (Spearman's rank correlation coefficient)
C16. Compute the map scale

## References

Anson RE, Ormeling FJ (1984–1996, (1st ed), 1993–2002 (2nd ed)) Basic cartography for students and technicians, 3 vols. Oxford: Pergamon Press and ICA; Butterworth

Anson RE, Ormeling FJ (1991) Basic cartography for students and technicians. Exercise manual. Elsevier Applied Science, Oxford

Claval P, Wieber JC (1969) La cartographie thématique comme méthode de recherche. Annales litteraires de l'Université de Besançon. Les Belles Lettres, Paris

PBNA cursus Cartografisch tekenen (1973–1976) Arnhem: Nederlandse Vereniging voor Kartografie and Koninklijke

Schulz G (1995) Lexikon zur Bestimmung der Geländeformen in Karten, 3rd edn, Berliner geographische Studien vol 28. Institut für Geographie der Technischen Universität Berlin, Berlin

Warn C (1980) The Ordnance Survey map skills book. Nelson, London

Kommission Aus- und Weiterbildung, Deutsche Gesellschaft für Kartographie e.V. (2000, 2004) Ausbildungsleitfaden Kartograph/Kartographin. München, DGfK

# Chapter 3
# The Role of Output Devices in the Higher Education Courses of Cartography

László Zentai

**Abstract** The rapid development of information technology affected cartography considerably in the last three to four decades of the twentieth century. In higher education curriculums the reproduction process was thought as a totally independent subject or a part of other related subjects. Cartographers should know the whole technical process to be sure that the maps will be reproduced as they were planned. As digital output devices become wide-spread and higher education institutes could afford these devices in the first phase the course about map reproduction was complemented by digital methods and the necessary theoretical information about the process was also added. In the second phase when the output device becomes easily available for even home users the traditional offset printing method has seemingly lost its importance and students were more interesting in digital printing methods. Nowadays one of the most important output devices in cartography is the screen of the computer, mobile phone, PDA which again made us to make changes in the curriculum and emphasize the importance of these new "output devices".

## 3.1 Introduction

The first independent cartographic courses were established in the Moscow State University of Geodesy and Cartography (MIIGAiK) in 1923. There were other important institutes concerning cartography at that time, in Zürich and in Vienna, but to establish an independent cartography course in these and other Western and Central European universities and high schools took longer time.

If we checking the curriculums of the early cartography courses we can realize the big difference between the old and modern structure. Actually there are much

---

L. Zentai (✉)
Department of Cartography and Geoinformatics, Eötvös Loránd University, Budapest, Hungary
e-mail: laszlo.zentai@elte.hu

L. Zentai and J.R. Nunez (eds.), *Maps for the Future*, Lecture Notes in Geoinformation and Cartography 5,
DOI 10.1007/978-3-642-19522-8_3, 

less courses offered in cartography than 30–40 years ago, most of these courses are converted to courses named as GIS, geoinformatics or geomatics. However the reproduction part of these courses (irrelevant to the title) has totally disappeared, while previously this area was an important part of cartography courses (Salichtchev 1979).

## 3.2 Traditional Reproduction Methods as an Output "Device"

The reproduction methods were always very important in map making. We can't simply treat these techniques scientific, there are mostly artistic or industrial methods. We don't have reproduction techniques which were developed directly for cartography, but cartography simply selected the best available methods that fitted their needs.

The printing methods of the first centuries after Gutenberg are not really important for cartographers (at least not for this paper). Most printing technology was based on letterpress, the printing of images that projected above nonprinting areas.

The first method we have to mention is a *lithography*. In 1798 *Alois Senefeler*, makes a discovery of profound significance in the history of artists' prints and later of commercial printing, too. He has been attempting for some while to print from limestone. What he comes to realize is that the antipathy between grease and water can be used as a basis for printing. He found that an image, no matter how detailed, that was drawn with a greasy substance on the face of a water-absorbent stone and then inked could be printed onto paper with absolute fidelity. Lithography was ideally suited for illustration (like maps) and enjoyed a phenomenal popularity during the nineteenth century, especially for color printing, which required a separate stone to print each color. The discovery of lithography was significant to the history and development of cartography. Prior to the birth of lithography at the turn of the nineteenth century, most maps and atlases were produced by engraving—a technique that requires much skill and labor. Engraved maps were rare and relatively expensive. Lithography offered a cheaper and quicker way to reproduce maps and other images. The early topographic map sheets were reproduced by engraving techniques, this method was suitable to reproduce the hachuring method of the relief representation.

Lithographic metal plates had only rarely been used for commercial printing, in part because the image on the plate was often worn through by the printing paper. In 1904 an American printer accidentally discovered that the lithographic image could be transferred, or offset, to a rubber cylinder that could then print as perfectly as the plate and would last indefinitely. This becomes the most popular printing process because of its economy, long plate life, and ability to print on many different textures.

The halftone color printing, the process still used today to reproduce full color, was introduced in the 1890s, but many years passed before its full potential was realized. Although color reproduction theory was fairly well understood, the lack of color film restricted color work to studios where the necessary separation negatives had to be made directly from the subject, under the most exacting conditions. Reliable color film became available in the 1930s and 1940s, and color reproduction grew both more common and more accurate.

The development of *offset printing* method was very important for cartographers. This was the first method which let the cartographers to reproduce their color maps economically without any constrain on the used colors or lines. However cartographers regularly used sport color printing methods instead of the process color methods which were the most common methods for color reproduction, especially for color photographs etc. In the early years of color offset printing it was not unusual that a map was printed more than ten spot colors which made the printing process quite expensive and required special care and treatment on the technology and the printing paper itself.

With the development of the photography screening methods (halftone dots) were used to reduce the number of printed colors using shades of the printed colors.

## 3.3 The Foundation of ICA

The *International Cartographic Association* was established in 1959. The foundation was closely connected with a new era of rapid and substantial development of cartographic technology. In the 10–15 years after World War II, against a background of a continuously increasing worldwide demand for maps (including not only topographic, but mostly road, city and tourist maps, but also thematic maps), an almost simultaneously occurring wave of innovations further revolutionized the map production process. Plastic drawing materials which were dimensionally stable were invented, new methods, like scribing on coated polyester material replaced the conventional ink drawing.

Typesetting machines were also very important in cartography because these machines could replace one of the most time consuming process of map production: the hand lettering. The first mechanical phototypesetters involved the adaptation of existing typesetters by replacing the metal matrices with matrices carrying the image of the letters and replacing the caster with a photographic unit. The industrial application of this idea resulted in the Fotosetter (1947), a phototypesetter manufactured in USA by Intertype. Very soon French, German and Russian models were invented. Later models with a separate keyboard printed more than 28,000 characters per hour. The third generation of phototypesetters appeared in the 1960s, in which all mechanical moving parts were eliminated by omitting the use of light and therefore omitting the moving optical device responsible for operating in its field (Latimer 1977).

*Dr. Carl Mannerfelt*, from the Esselte Map Service (Sweden) initiated a cooperation of map production experts. He invited a number of foreign experts in different areas of map production to exchange information on the technological innovations (including map editing, compilation and reproduction). It was in 1956 and due to the Cold War period only Western countries were invited. The main reason of the success of this first meeting was that the participants were focused on only special areas of cartography: the map production, there were not scientist; they were practical cartographers, map producers. When ICA was formed the scientific aspects became more important (international co-operation in the field of cartography), than the technical aspects (according to the final resolution of the Esselte Conference on Applied Cartography the planned international organization should concentrate on such aspects of cartography as are not already covered by existing organizations, like IGU, FIG, etc.).

On the first ICA General Assembly in 1961 twenty-six countries were inaugurated as member countries. The Statutes as adopted in 1961 did not represent either governmental or commercial cartographic interest. Its aims were the study of cartographic problems, the co-ordination of cartographic research involving co-operation between different nations, the exchange of ideas and documents (and later digital data), the training for cartographers and encourage the spreading of the cartographic knowledge. However in the first years of ICA when the relationships between IGU and ICA was a part of a long discussion it was said that cartography as a technical science might be subject to commercial and governmental influences. It was really a fact that the Esselte Conference on Applied Cartography in 1956 and the other early meetings were initiated by private map producing companies so it was probably the main reason that socialist countries were not invited and started to become ICA members only after 1964 (nevertheless it was also a political issue at that time).

The definition of the term cartography was also an important task of the early years of ICA (Stéphane de Brommer, the Vice-President of ICA and the chairman of the Commission on Education and Training). The Multilingual Dictionary of Technical Terms in Cartography (edited by Emil Meynen in 1973) declared the term as "the art, science and technology of making maps, together with their study as scientific documents and works of art". This definition was a compromise after a long discussion inside the association, but Prof. Konstantin Salichtchev, the ICA President (1968–1972) still regretted that the final definition was concentrated too much upon the map production and didn't concentrate enough on a scientific approach. The definition later has been changed due to the computer technology, but this issue still shows the importance of the output in cartography (Ormeling 1988; Salichtchev 1979).

### *3.3.1 ICA Commission on Map Production*

The initial statutes of the ICA only roughly described as to how the commissions should be established. The first ICA commission on focusing partly on the reproduction methods was the Special Commission on Automation. This commission was established in 1964 and in 1980 it was renamed to the Commission on Computer-Assisted Cartography. The commission had very general terms of reference which later became more specific including the development of cartographic systems (whatever it means). The only objection to the commission's work was that it was usually highly based on the hardware elements of computer cartography (including output devices), but the cartographic presentation was often hidden in the background. This opinion on computer-assisted cartography was still valid in the 1980s. The series of AutoCarto conferences (USA) had a strong influence on the technical development of computer cartography, but the results of this dynamic process were used only in the most developed countries.

In 1971–1972 a questionnaire result demonstrated that ICA member countries thought that there was a need for a commission to study practical map production methods and techniques. The Commission on Cartographic Technology was established in 1972 with the following terms of references:

- To review current cartographic techniques and processes.
- To disseminate information on these techniques and processes.
- To organize commission meetings.

In order to avoid overlapping with the above mentioned Special Commission on Automation the new commission restricts its activities to study

- Color proofing techniques (which is an important part of the map production process);
- Register systems (these systems were developed to make the offset printing process smoother and to avoid mis-registering);
- An evaluation of techniques for the production of small editions of multicolored maps (this problem is still existing, but the digital printing methods are now easily available, which was a not a case when the commission was originally formed).

The commission was very active in the early years, they continuously monitored the new development of the map production technology, but due to the production of different written manuals and handbooks their activities were partly slowed down. The name of the commission changed several times and now it is called Commission on Management and Economics of Map Production. Although the subjects of the commission has changed comparing to the early years, but matters like printing-on-demand, web publishing and archiving should be researched by the commission since the middle of 1990s.

In 1984 the first volume of *Basic Cartography* for students and technicians was published (Fig. 3.1). All together three volumes and an Exercise Manual was published. The Volume 3 was published in 1996 and dealt mostly only with computer cartography. But in the first two volumes there was all together 10 large chapters. The Map reproduction chapter was written by Christer Palm in Volume 1 (Sjef van der Steen was a co-author in the second edition), but additional chapters like Techniques of map drawing and lettering, Map compilation and Computer-assisted cartography have also partly dealt (or at least mentioned) to the output part of cartography. The effect of the Basic Cartography volumes was important since the original formulation of the concept of the series cartography,

**Fig. 3.1** Cover page of the basic cartography volume 1 (first edition)

like many other fields of activity within the general area of information technology, has undergone rapid changes in both form and character. Basic Cartography had an important role also in the higher education to make an internationally standardized teaching material for cartographers. However the rapid development of computer cartography has made difficult and complicated to integrate the knowledge represented in these volumes into the curriculums. In less developed countries where the use of computer technologies was late due to the limited financial resources Basic Cartography played an important role to represent an international standard of cartography. The Map reproduction chapter mentions only the traditional printing methods (offset print), but the chapter deals with other technical subprocesses (Ormeling 1988).

## 3.4 Traditional Reproduction Methods in the Cartographic Education

Cartographers had to understand the printing process, so the basics of the offset printing process were part of the higher education cartography curriculums.

*Erwin Raisz*, one of the most important American cartographers of the twentieth century felt that cartographers fell into two categories: "geographer cartographers," who wish to express their ideas with graphs, charts, maps, globes, models; and "cartotechnicians," who "help produce maps, models, and globes by doing the color-separation" or other technical works. He proposed the idea of different types of cartographers, including the cartologist, cartosophist, toponymist, map compiler, map designer, draftsman, letterist, engravers, map printers, etc.

Of course, it may take time to establish independent cartography courses and before the World War II practically these independent courses didn't exist. The other important American cartographer of the twentieth century was Arthur Robinson who got his PhD right after the World War II in the Ohio State University. On that university subject like Cartographic Production was part of the course. Raisz's General cartography and Robinson's Element of Cartography are probably the most important books in the twentieth century of the American cartography, but we should mention that the cartographic education was most developed in Europe at that time.

In the Latin American and European traditions, the production techniques are certainly not considered part of the academic environment. They are very respected and much appreciated, but generally we will find them not in academia (McMaster and McMaster 2002).

Although without fully understanding the essence of offset printing cartographers were not able to create a good symbology (which can be reproduced technically), they have to start the whole process with the definition of the printed colors, line widths etc. To increase the number of printed colors may considerable make the whole map production process more expensive (it was even more relevant around 1960–1970). Even the selection of the number of color shades was a question of costs. It was very important in the education of cartographers (especially in the higher education where we have time to give complex knowledge and make the students understood the relationships) to clearly understand how the final phase of the map production process (offset printing) may affect the beginning phase of the map production (creating a map symbology). However it is not obligatory to be an expert of the whole reproduction phase, but to understand its importance is really necessary if we want to run our business efficiently.

In my university the independent Department of Cartography was founded in 1953 and in 1955 we started the first cartography course. In practice it was not an independent cartography course, but it was a specialization after the second year based on the geography and geology courses (we may call it is a pre-Bologna system: 2 + 3 years). We had a series of lessons with a title Map technologies and the students had to understand the whole technical process of map producing: map drawing or scribing, all kind of technical works (photography, screening), proofing, offset printing (including binding, folding). These were not just theoretical lessons, but students had to practice all different areas: we had all technical devices (special camera for reproduction, frames for contact copies, special instruments for screenings, proofing devices, offset printing machine; folding and binding were connected to book and newspaper publishing so we didn't have such kind of machines so we visited printing houses to present these processes to our students). Of course we also had similar practical lessons in the drawing part (ink drawing and scribing, etc.). A student who has made the whole process himself/herself was able to understand the potential errors so they were able to manage the production line when they started to work in mapping company or state cartography.

## 3.5 The First Steps of Digital Cartography

The *Geographic Information System* was parallelly developed in America, Australia and Europe. The extraction of simple measures largely drove the development of the first real GIS, the Canada Geographic Information System (CGIS) in the mid-1960s. CGIS was planned and developed as a measuring tool (mostly to help the precise measures of areas), a producer of tabular information. At that time the system couldn't be a mapping tool, because the most important input and devices were not yet developed. At that time computer systems were very unique and very expensive and mainly used only for research or (secret) military purposes.

A second step of innovation occurred on the late 1960s in the US Bureau of the Census, in planning the tools needed to conducts the 1970 Census. The DIME program (Dual Independent Map Encoding) created digital records of all US streets, to support automatic referencing and calculation of census records (the US Census was always very innovative, Hollerith built machines under contract for the Census Office, which used them to tabulate the 1890 census data using special punched cards). The similarity of this technology to that of CGIS was recognized immediately by experts and led to a major program at Harvard University's Laboratory for Computer Graphics and Spatial Analysis to develop a general-purpose GIS. Early GIS developers recognized that the same basic needs were present in many different application areas, from resource management to the census.

In a largely separate development during the second half of the 1960s, cartographers and mapping agencies had begun to ask whether computers might be adapted to their needs, and possibly to make the map producing more cost effective. The UK Experimental Cartography Unit (ECU) pioneered high-quality computer mapping in 1968; it published the world's probably first computer-made map in a regular series in 1973 with the British Geological Survey; the ECU also pioneered GIS work in education, and much else. National mapping agencies, such as Britain's Ordnance Survey, France's Institut Geographique National, and the US Geological Survey and the Defense Mapping Agency began to investigate the use of computers to support the editing of maps, to avoid the expensive and slow process of hand correction and redrafting. The first automated cartography developments occurred in the 1960s, and by the late 1970s most major cartographic agencies in the Western part of the World were already computerized to some degree. But the magnitude of the task ensured that it was not until 1995 that the first countries achieved complete digital map coverage in a database (including digital state cadastral and topographic maps series) (Longley 2005).

*Remote sensing* also played an important part in the development of GIS (and also cartography), as a source of technology and more importantly as a source of data. The first military satellites of the 1950s were developed in great secrecy to gather intelligence, but the declassification of much of this material in recent years has provided interesting insights into the role played by the military and intelligence communities in the development of GIS. Although the early spy satellites used conventional film cameras to record images, digital remote sensing began to replace them in the early 1970s. At that time civilian remote sensing systems such as Landsat were beginning to provide vast new data resources and to exploit the technologies of image classification and pattern recognition that had been developed earlier for military applications. Weather satellite images had also an important role especially in meteorology where the low resolution of the early images was good enough to improve the precision of global weather forecasts. Nowadays satellite remote sensing provides the best quality large-area coverage database on Earth (Harris 1987).

Although there is no close contact to the output side of the cartographic process we also have to mention another important step of digital cartography, the global satellite navigation systems such as GPS. However GPS became widely used in the civil cartography only in the twenty-first century, although for military use it was already available around 1980.

## 3.6 Computer Printers

There is no space to give a comprehensive overview of computer printing in this paper, so only the most relevant techniques will be discussed. The development of computer technology before the release of personal computers was focused on the calculation speed. This was also the time when the theory of the geographic information systems was developed. In the 1960–1970 years there was not too much focus on output devices in informatics. At that time matrix printers were nearly the only opportunities to make output from digitally stored data. These devices were produced to print text (characters) based on simulating the well-known "output device", the typewriter using a so called impact printing. However these devices were driven by the computer at a much larger speed than any human could do. What was much more important in the cartographic point of view is the print size. In the industrial and scientific environment where financial resources were available such kind of large format line matrix printers and (later) dot matrix printers were used. This large output size (wider than the 80 columns of an A4 sheet) became available in the 1970s. The Harvard Laboratory for Computer Graphics (and Spatial Analysis) developed an automated mapping application called SYMAP, to produce isoline, choropleth and proximal maps on a line printer around 1960. This technology was used only in research institutes where line printers were available (Fig. 3.2). This printing technology has never been used together with personal computers, however the technology was long time used in business environment

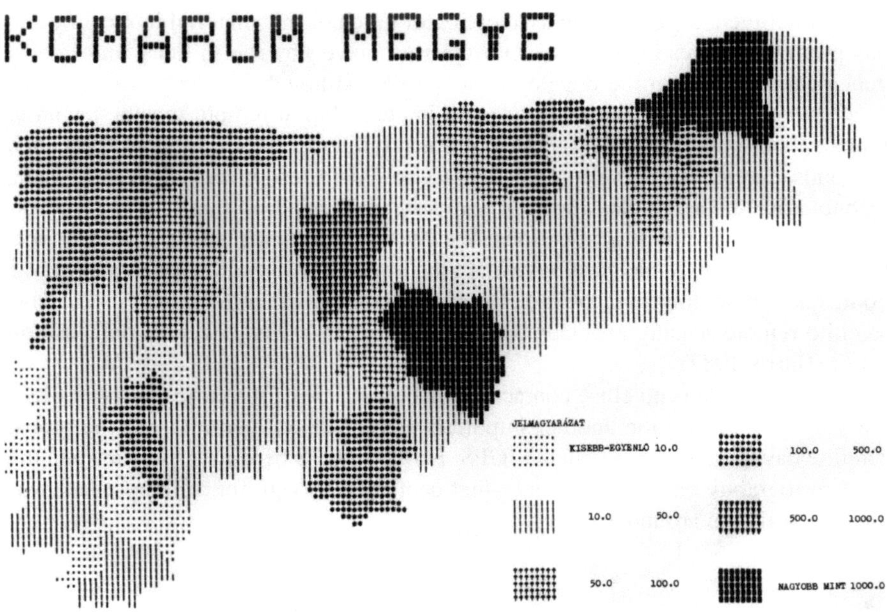


**Fig. 3.2** Hungarian "SYMAP style" thematic map, Komárom county (made around 1975)

(bulk printing), but finally the large speed laser printing replaced the old technology. This technology probably didn't appear at that time in the cartographic education, but nowadays it is widely thought when we want to present the evolution of digital cartography and geographic information system.

Dot matrix printers were developed around 1970. The very first printers used 5*7 dot matrix to form the characters, later the 9*9 dot matrix became standard. In the 1970s and 1980s, dot matrix impact printers were generally considered the best combination of expense and versatility, and until the 1990s they were by far the most common form of printer used with personal computers. Early dot matrix printers were notoriously loud during operation, a result of the typewriter-like mechanism in the print head and they produced printouts of a distinctive "computerized" quality (the quality of text printing was far from the real typewriter text). Although the very first dot-matrix impact printers lacked the ability to print computer-generated images it has changed soon and encouraged the PC users to buy this relatively cheap output device. The speed and graphic quality was very poor, but users had no other opportunity to print their documents, including maps. This was also the first low-cost option for color printing (although they were also 9 pin dot color models). When the manufacturers wanted to improve the printing quality of their models they invented a 24 pin dot matrix printer around 1985. The print quality was not comparable for the actual models, but these were the first printers allow the users to print color photographs (even such prints might take some 10 min or more even in a small size). The color dot matrix printers had no chance to become wide spread because a new technology, the inkjet was invented and in some years this technology replaced the dot matrix printers especially on the home market. The effect of color dot matrix printers in cartography was nearly invisible due to the fact that only few manufacturers developed such models (Apple, Citizen, Epson, Panasonic), and the print quality was really poor (printing stripes remained visible on the paper) (Fig. 3.3). Some color dot matrix printer models are still on the market, but only for printing receipts (Zable and Lee 1997).

The very first special output devices used in cartography was the pen plotter. Compared to modern printers, pen plotters were very slow and cumbersome to use. Users had to constantly worry about a pen running out of ink. If one pen ran dry at the end of a plot, the entire plot had to be re-done which was very time consuming. Plotters could only draw lines; they couldn't reproduce raster or photographic images. Despite these limitations, the high resolution and color capability of pen plotters made them the color hardcopy output device of choice until the late 1980s especially in technical drawings and CAD graphics which consisted on simple lines like cadastral maps. Only the inkjet technology has made pen plotters obsolete.

In a drum technology papers were fixed and pen is moved in a single axis track and the paper itself moves on a cylindrical drum to add the other axis or dimension. Where the paper was fixed on a flat surface and pens are moved to draw the image was called a flatbed plotter. This type of plotters regularly can use several different color pens to draw with.

*CalComp* was incorporated in 1958, it was one of the first companies in the United States to market peripheral products designed specifically to work with

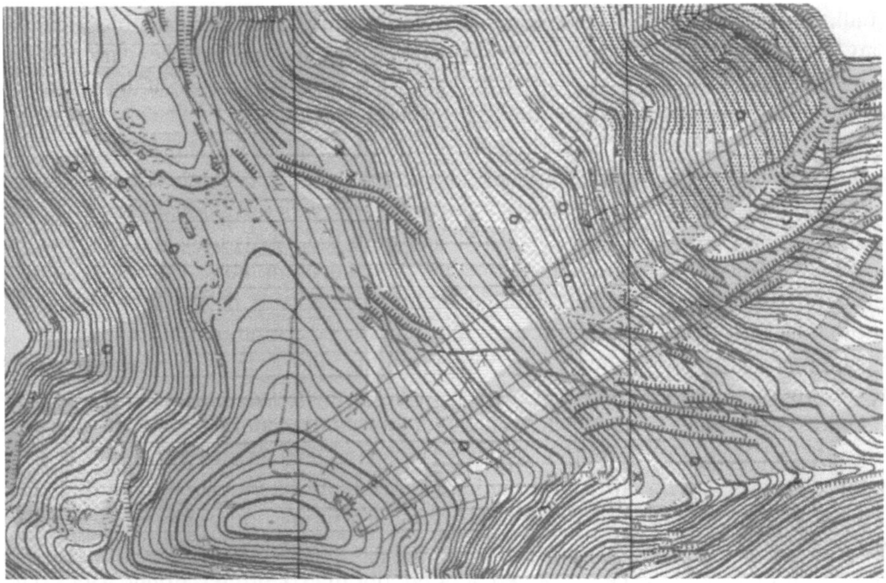

**Fig. 3.3** Twenty-four pin dot color matrix printed map

computers. CalComp's appearance coincided with the first wave of acceptance of computers by such mainstream businesses as banks and insurance companies. In 1959 the company developed the world's first drum plotter, but few expected the instrument to grow into CalComp's strongest product line (Fig. 3.4). The company introduced a complete line of drum plotters in 1962. By 1968 between 80% and 90% of all plotters in existence were manufactured by CalComp (prices were between $3,500 and $50,000).

The other important plotter producing company of the early years was Versatec. They produced the first commercially successful electrostatic writing technique plotter that produced information on paper directly from digital data sources in 1970.

Hewlett Packard and Tektronix produced small, desktop-sized flatbed plotters in the late 1960s and 1970s. The pens were mounted on a traveling bar, whereby the y-axis was represented by motion up and down the length of the bar and the x-axis was represented by motion of the bar back and forth across the plotting table. Due to the mass of the bar, these plotters operated relatively slowly. HP produced only flatbed plotters before 1980, but the model 7580A (released in 1981) was the world's first "grit wheel" pen plotter. This machine combined high speed and high line quality in a small package at a price less than half that of comparable products on the market at the time. Small grit-covered wheels move the paper along the X-axis of the 7580A, thereby replacing the heavy, bulky components used in other plotters with a low-mass, low-inertia drive mechanism.

**Fig. 3.4** CalComp model 565, a 12 in. drum plotter (1959)

Plotters were perfect devices in the early years of digital cartography, but their high price has made these devices affordable only for large companies. When the price was reduced these plotters were also used in the higher education because the programming was relatively easy (very simple instructions on the plotter graphic languages) and students can use it efficiently. It was also important in the higher education that the operational method of the plotter was theoretically easily understandable. The operational costs very relatively low (pen, paper) and the devices didn't require special treatment which was important in the higher education.

On our department the first real digital output device was a simple *Numonics* drum pen plotter which we acquired in the beginning of 1990s and our students printed different projection coordinate networks which was driven by the students' computer programs.

## 3.7 Actual Computer Printing Technologies

Since the graphic operation systems and graphic environments were released around 1984–1990 and graphic software also become available computer printers started to be a part of a cartographic workstation. Laser printers were fast and reliable, but color models were very expensive in the beginning. This technology doesn't allow larger models than A3 size. The prices were falling down in the last some years so laser printers (including color ones) are real options now even in the home market.

Inkjet technology replaced the dot matrix printing in the middle of 1990s (Canon released their very first model in 1981 and HP did the same in 1984). Although black and white models were less expensive in beginning, but the price difference between color and black and white models was not very high. In some years black and white models have totally disappeared. This was the first technology that reached the photo quality on a reasonable price. However photo quality and color accuracy was not a must for cartographers, but it helped to improve the technology (for example using more than four inkjet cartridges to be able to reproduce more colors). Before manufacturers have reached this goal they just increased the resolution of their models up to 1,200 dpi. However the photo quality was not only the issue of resolution, but also the quality of ink and paper.

The large format models were much more important for cartographers, although primarily the large format models were developed for CAD. This was the reason that the first large format models were concentrated on line drawing instead of filling area with homogenous colors.

Other technologies like dye sublimation, thermal vax, electrostatic printing were so expensive that only state cartography institutes were able to afford that, but the operational costs were very high so inkjet models replaced these special printers in some years. These special technology models are still available and they used where color accuracy is a key factor, but color management is also available for inkjet models.

Inkjet printers appeared soon in the higher education. A4/A3 models were available on a reasonable price already in the first part of the 1990s and the price range of large format models was also reachable even for higher education department. Even students could afford such printers on the second half of 1990s (Zentai 2009).

### *3.7.1 Computer Printing in the Education*

Only when the digital printing become an affordable option in the higher education it was really important to teach not only the cartographic software, but also the digital printing process. Although the first printers were not really sophisticated, but their appearance gave the users a freedom, not to depend on the expensive, time consuming process of offset printing. In the beginning digital printing has just treated as an additional proofing method, which has substituted neither the offset print, nor the proof. The real advantage of this new option was the low cost and the speed. The speed was really slow comparing the actual capabilities, but it was much faster than any contemporary analogue method.

Students were very interesting on the digital technologies (as younger generation regularly did so) so they wanted to use these technologies as early as the higher education institutes can provide them the opportunity. But it was also important to teach the theoretical background of the digital printing process which has considerable differences comparing the traditional offset printing. One of the most important differences was that color digital printing used a CMYK color model instead of using spot colors which was very common in the traditional map printing. This difference is also influenced the composition of map symbols (colors, line widths, etc.) and the early years of the digital era we have to take into account the weakness of the technology. It was very important to make students understood these characteristics of the printing process.

As the price of computer printers decreased these devices become part of the home computer systems together with scanners. The internet era has made the access of information (especially the IT information) much easier than any previous time. Such kind of knowledge was partly thought in secondary schools, but it is still necessary to teach the theoretical and technical background for the cartographers. It is similar to the offset printing era: some cartographers become the expert of the technology and they were responsible for that part of the process. Nowadays we also have only few cartographer experts who really understand the process and the rest (as any other user) is just press the print button and trust in the software and hardware.

It is also necessary to mention the new output device: the screen. More and more maps are planned (or at least seen) on the screen of computers, mobile phones, GPS devices, personal digital assistants. To visualize the digital information efficiently students have to be familiar the characteristics of these new “devices”, so such kind of subjects should be included in the curriculum.

## 3.8 Summary

As cartography become science and independent cartography courses and curriculum were implemented in the twentieth century the content of these studies has been developed continuously. After the International Cartographic Association was formed they formed commissions to encourage the cooperation of the scientists and higher education experts of the member countries. The ICA Commission on Map Production was established in 1964 and played an important role in developing, standardizing the modern reproduction techniques of cartography. The teaching of map production techniques in the higher education was also influenced by the commission, especially in less developed countries, but as the computer printers become widely used this effect started to decrease. Cartography curriculums tried to follow the development of these output devices and even today when digital maps are the most important products of cartography we have to teach the visualization on computer screens. But the development is continuous: 3D screens and other devices are coming, making new challenges also for cartographers.

**Acknowledgement** The European Union and the European Social Fund have provided financial support to the project under the grant agreement no. TÁMOP 4.2.1./B-09/KMR-2010-0003.

## References

Harris R (1987) Satellite remote sensing. Taylor & Francis, London
Latimer HC (1977) Preparing art and camera copy for printing: contemporary procedures and techniques for mechanicals and related copy. McGraw-Hill, New Work
Longley P (2005) Geographic information systems and science. Wiley, New York
McMaster R, McMaster S (2002) A history of twentieth-century American academic cartography. Cartogr Geogr Inf Sci 29:305–321
Ormeling FJ (1988) ICA 1959–1984. The first twenty-five years of the International Cartographic Association. Elsevier Science, Enschede
Salichtchev KA (1979) Periodical and serial publications on cartography. Can Cartogr 16:109–132
Zable JL, Lee HC (1997) An overview of impact printing. IBM J Res Dev 41:651–668
Zentai L (2009) Change of the meaning of the term 'cartographer' in the last decades. In: XXIV international cartographic conference, Santiago de Chile, 15–22 Nov 2009. ICA

# Chapter 4
# Teaching Cartographical Skills in Different Educational Systems of EU

**Eszter Simonné Dombóvári**

**Abstract** The European Union means a union of countries with different past and development, various national traditions, customs and educational systems. The main aim of these educational systems is to import growing and basic knowledge in a socially institutionalized way. Teaching cartography is one little piece of these systems and it appears at different level of these education systems. We can obtain this knowledge in many ways with the help of numberless educational tools. With the new possibilities of the web mapping 2.0, method and content of teaching expands and a new quality in teaching and training comes into life. The aim of this research is to find the answer on the questions what the role of cartography knowledge is in various educational systems; how different educational systems influence teaching cartography; how we can improve the cartographic education in schools with experiences from different countries; what the role of new media in education is and beside traditional tools and what new techniques are offered in the field of cartography teaching. These experiences of comparisons can help us to improve cartographic education in schools.

## 4.1 Introduction

The European Union means a union of countries with different past and development, various educational systems, national traditions and customs. The member states pay great attention to care for cultural diversity and preservation of national values (Ormándi 2006). These differences give the multicolourness of the Union.

The European Union was an economic cooperation until the 1990s but it is a lot more now, a global economic and political integration which also affects education. A kind of globalization can be observed in this field as well. The role of the

E. Simonné Dombóvári (✉)
Institute of Geoinformation and Cartography, Vienna University of Technology, Vienna, Austria
e-mail: eszter@cartography.tuwien.ac.at

L. Zentai and J.R. Nunez (eds.), *Maps for the Future*,
Lecture Notes in Geoinformation and Cartography 5,
DOI 10.1007/978-3-642-19522-8_4, 

**Table 4.1** Politics of education in EU (Ormándi 2006; Kőfalvi 2006; OKM 2010; OKI 2000)

| Date | Events |
|---|---|
| 1957 | Treaty of Rome: European Economic Community (EEC) was created. Decisions affected training |
| In the 1970s | The importance of educational cooperation grew between Member States |
| | Culture and Education Commission of the European Parliament and the European Union Council of Education, Youth and Culture Council (the Council of Ministers of Education) |
| In the 1980s | European cooperation and mobility (Erasmus, Petra, Commett, Lingua) |
| In the 1990s | 1992 the Maastricht Treaty: clarification of educational powers |
| | 1996 European Year of Lifelong Learning |
| | 1993 Green Paper, 1995 White Paper, 1997 Blue Paper on the European Dimension of Education |
| After 2000 | 2000 Lisbon: Education and Training 2010, establishing the European Qualifications Framework (EQF) |
| | 2003 program for the effective integration of information and communication technology (ICT) in European education and training systems (2004–2006)—European Parliament and Council Decision |
| | 2006 recommendation for lifelong learning, key competences |
| | 2007–2013 Lifelong Learning Programme |
| | 2008 recommendation for lifelong learning in the European Qualifications Framework (EQF) |
| | 2009 the year of creativity and innovation |
| | 2010 work program of Education and Training |

community is growing. In the interest of faster development, the inner educational politics of the member countries are coordinated. As standards unify, it needs unifying teaching that leads to unified compulsory education (Kozma 1997).

The competitive and continuous knowledge renewal, the role of skills and competencies appreciate which is the primary source of education and training (OKM 2009). Nowadays the goal is an education and a training system of high quality (Table 4.1). A competitive and knowledge-based society has brought "lifelong learning" as well as the competency-based education (appropriate for student's skills) for the last decade, and also has posed challenges to the traditional educational systems.

The EU recommendations deal with education, training, adult learning, lifelong learning, mobility and e-learning (OKM 2010). Nowadays the member states decide only on their own educational politics (compiling teaching material as well as their school systems). Their task is the implementing of the European Union programs and they are responsible to revamp their education system. The Union is responsible for promoting cooperation, supporting the process and encouraging additional community programs to achieve a competitive union (Ormándi 2006).

The differences of the educational systems enhance the role of international comparisons. The base of the comparison may be e.g. the length of compulsory education and of teaching periods, the structure of pre-school, primary, secondary and tertiary education and educational means (Ormándi 2006). My aim of the resource is to show the educational systems and teaching of geographical and cartographical knowledge in a few member states of the EU.

## 4.2 Educational Systems in European Union

One of the most important tasks of public education is knowledge transfer based on educational systems with several different layers of learning and training. The *educational systems* of a given country include all the schools of the country and show an organization where schools (of different type, level, function, maintenance) are connected horizontally and vertically (Mezei and Szebenyi 1998). Its main aim is to import growing and basic knowledge in a socially institutionalized way.

Every single school system has evolved and developed differently in the influence of different factors (e.g. traditions, historical, social and economic development) shaping them. Therefore there are similarities and differences between them. Differences may not only be in maintaining schools but also in education—such as the length of compulsory education, the age of school beginning, different school types, school system and interoperability between schools, the level of state influence.

To examine every single different method we need a homogeneous scheme. UNESCO, OECD, EU have formed *international standard classification of education (ISCED)* in order to interpret educational training forms and programs homogeneously (UNESCO 1997). With its help the different educational systems can be compared easily. There are seven levels determined within educational systems (Fig. 4.1).

The *Eurydice Network* contains information and analysis of educational systems and policies in 31 of the members of the EU Lifelong Learning Programme in Europe: the 27 EU countries, Liechtenstein, Norway and Iceland (as member of the European Economic Area, EEA) and Turkey. The publications and the databases of the member states' education and its comparison are organised by the EU Education, Audiovisual and Culture Executive Agency in Brussels (Eurydice 2011) which formed the basis for this comparison.

The compulsory education lasts 9–10 years in most EU states (in Luxembourg, Malta and UK 11 years, in the Netherlands and North Ireland 12 years and in

| | |
|---|---|
| Level 6 | Second stage of tertiary education (leading to an advanced research qualification) |
| Level 5 | First stage of tertiary education (not leading directly to an advanced research qualification) |
| Level 4 | Post-secondary non-tertiary education |
| Level 3 | (Upper) secondary education |
| Level 2 | Lower secondary (second stage of basic education) |
| Level 1 | Primary education (first stage of basic education) |
| Level 0 | Pre-primary education |

**Fig. 4.1** ISCED levels

Hungary 13 years). It is between the ages of 6–16. In Luxembourg, Ireland and the Netherlands compulsory education starts at the age of 4, while in Bulgaria, Estonia, Finland, Lithuania, Denmark and Sweden only at the age of 7. Usually the start of the compulsory education coincides with the start of the primary education. In case of earlier start, the pre-school program is part of the primary education. Primary education is generally completed at the age of 12, but it is linked up to lower secondary education in some countries and it can take to the age of 15–16 and the upper secondary education to the age of 18–19.

In most countries the school pathways for pupils are generally identical up to the end of the lower secondary level (14 or 15 of age). Malta, Poland and the United Kingdom have core curriculum up to the age of 16. In some countries, pupils can choose a specific type of school at the beginning of lower secondary education (Germany, Austria, the Netherlands and Luxembourg). In other countries compulsory general education is organized in single-structure schools up to the age of 14 or 15 without a transition between primary level and lower secondary levels. But from the age of 10 or 11, pupils in some countries can attend another school—Czech Republic, Latvia, Hungary and Slovakia (Herodot 2007).

Different levels have developed differently in the countries of the European Union, but structural similarities can be recognized (Figs. 4.2, 4.3). We can distinguish three models where different levels—elementary, lower and upper secondary—are connected or separated (Lannert and Mártonfi 2003):

- *The model with three cycles (elementary + lower secondary + upper secondary)* can be 5 + 4 + 3 (e.g. in Italy) or 6 + 3 + 3 (e.g. in France).
- *The model with two* cycles *(elementary and lower secondary + upper secondary)* where the first period is longer and the second one is shorter. It can be 8 + 4, 9 + 3, 9 + 4. The countries are e.g. Portugal, Scandinavian countries and countries in East-Europe.
- *The model with two cycles (elementary + lower and upper secondary)* has a 4–6 year first period and a 6–9 year second period (4 + 8, 6 + 6, 4 + 9) e.g. in Austria, Germany, Belgium, The Netherlands, Ireland etc.

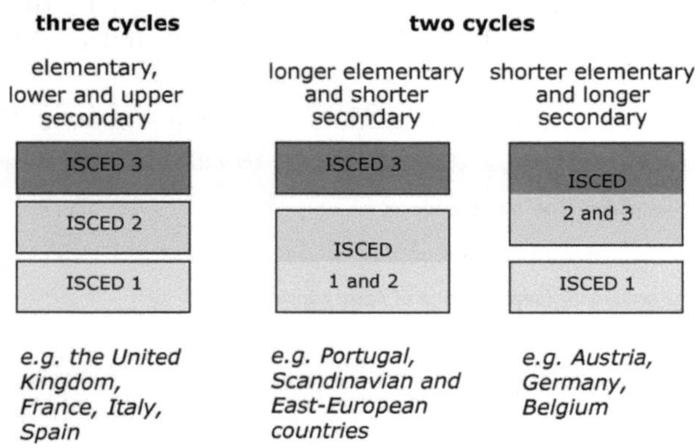


**Fig. 4.2** Types of school systems in the European Union (with ISCED levels)

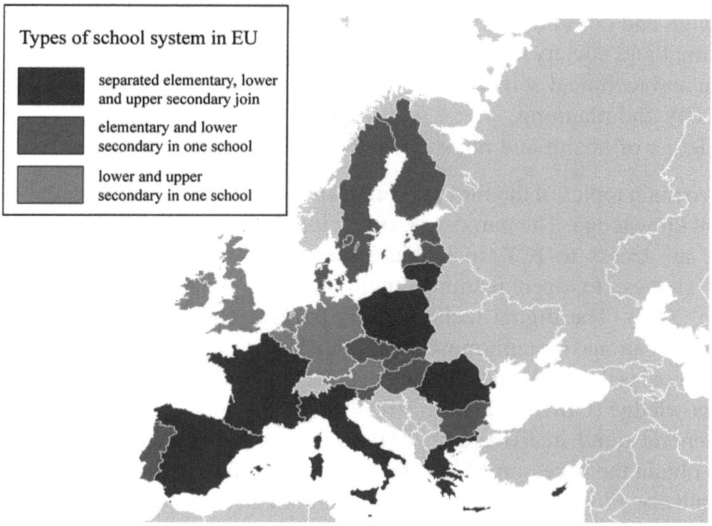


**Fig. 4.3** Types of school system in the European Union

Different models influence learners in different ways. The model with three periods in different schools means more changes for children. When the first period is short, selection is too early. (The early selection process means, the children of the largest segment of the population are underrepresented at higher schools. Educational systems tend to reproduce traditional social structures instead of being a vehicle of opportunity or social mobility). If the first period is longer, foundation of teaching is better. For the last years, emphasis is on openness and individualization of education, on the process of teaching and learning, on developing claim for learning, on improving learning abilities, on attaining active learning strategies, on forming abilities to cooperative learning. Learning has expanded both in space and time and it takes place more and more outside school building. These changes need a foundation of good quality where formal schooling is in dominant position (Lannert and Mártonfi 2003).

## 4.3 Curriculum in European Union

The *curriculum* is a partly pedagogical, partly educational guidance document which consists of general teaching and training aims, aids of didactics, offered teaching method, lesson plans, teaching material for subjects and grades (OKM 2003). The aim of the curriculum for each school type is to develop subject-specific and new competences (new competences: e.g. foreign languages, entrepreneurship, social skills and technical culture, information and communication technology—ICT). There are five separate areas of studying:

1. Language and communication,
2. Individual and society,
3. Nature and technical sciences,
4. Creativity and planning,
5. Knowledge of health and physical education (Eurydice 2002).

The two main topics of the European educational policies are the foreign languages and digital knowledge. The purpose is to develop skills to a knowledge-based society, to ensure an access to ICT, to increase the number of students in scientific and technical studies, to open learning environment, to make teach more attractive (Eurydice 2002). The digital technology is transforming every aspect of people's lives. The educational systems must adapt to these individual needs and requirements. Today, the goal is the skills and the competencies required for practical life rather than factual knowledge (Ormándi 2006).

The general trend in EU member states is keeping students for a long time in school to obtain the necessary skills and to prepare "lifelong learning". The period of the compulsory education increases. The time devoted to nature and social science subjects as well as foreign language education is growing. The following disciplines are emphasized in the different stages of the educational systems (Eurydice 2009):

- The curriculum of *primary education* is almost the same in all member states. The differences are only in flexible timetables and obligation to provide ICT instruction and religious or moral instruction. The main subjects are the language of instruction (one-quarter and one-third of teaching time), mathematics, sciences and social sciences (9–15% of teaching time), physical education (7–12%), foreign language (less than 10%), religious and moral instruction (4–8%). The information and communication technology (ICT) education as a subject takes very little time and often belongs to other disciplines.
- The timetable of the subjects is changing in the *lower secondary school*: almost every member state reduced the proportion of the native language and mathematics, respectively increased the proportion of natural and social sciences. The natural science will find the largest number of lessons in some countries (the Czech Republic, Estonia, Slovenia, Slovakia and Finland). 10–20% consists of teaching foreign languages, mother tongue, mathematics, natural and social sciences. Foreign language education is the same ratio compared to the primary education but the education of arts is less at this level.
- In *upper secondary education* general education and vocational education are separated.

### *4.3.1 Teaching Geography*

*Teaching Geography* in each member state can be characterized by autonomy in the subject, changes in content, attitude and number of lessons (Fig. 4.4). *Teaching Cartography* is one little piece of geography education but this is one of the most

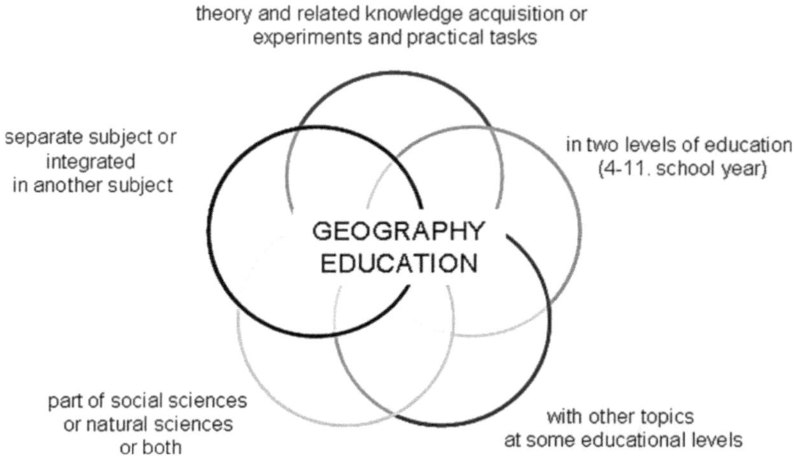


**Fig. 4.4** Several aspects of teaching geography in EU

important knowledge to show the spatial and temporal trends of geographical phenomena and processes. The geographical content is an important part of the curricula, such as cartography. This is a compulsory curriculum either it is a separate subject or it is integrated in another subject. It is taught at least in 2 years between level 4 and 11 (Curić et al. 2007).

Geography is required for different long time in each member states. In most countries it is a compulsory subject between the age of 15–16 (Austria, Czech Republic, Denmark, Finland, Germany, Hungary, Ireland, Italy, Latvia, Lithuania, the Netherlands, Poland, Portugal, Spain and Sweden). In Cyprus, Greece, Italy, Slovenia and the United Kingdom it is the age of 13–14, in Belgium, Malta and Slovakia from age of 10–12, but in Bulgaria, Estonia, France, Poland and Romania, it must be studied to final exam of the secondary school (Herodot 2007).

The role and the teaching time of geography has been reduced both in primary and secondary education. Some progress can be observed in Estonia, Slovenia and Sweden (Herodot 2007). The decrease of the lesson number is observed in all countries. Lessons per week: 7 h per week in Hungary (3 + 4, primary and secondary education), Slovakia (16), Slovenia (12), Romania (12–17, depends on school), Poland (12) and Austria (16) (Ütőné 2009). But in many cases only the curriculum determines the teaching time of individual subjects and each school can decide the exact curriculum: e.g.

- In Sweden the government provides the teaching time of the subjects for the duration of 9-year-education.
- In Ireland there is the minimum amount of time.
- In the Netherlands schools have the largest autonomy. Schools themselves decide the number of lessons.

- In England the teachers can decide lesson number, but the theme should be tailored to the curriculum.

In some countries the theory and related knowledge acquisition is important, but less attention is paid to practical cognition. Such are the countries in Central and Eastern Europe, e.g. Czech Republic, Slovakia and Hungary, which have a so-called traditional way of science teaching. However, the Anglo-Saxon and Nordic countries have a major role in the experiments and practical tasks. In the latter countries, science or integrated science courses appear in general (TIMSS 2007).

In every country: the preparation of educational content, the use of different teaching methods and forms, different types of student work (individual work, testing, pair and group work, discussion, field work, project, presentations, essays, posters, projects, presentations) is important. Furthermore, it is important to connect geographical knowledge to other subjects.

In the 1990s, Geography, as a subject, began to be independent. It is an integrated subject in primary education in most cases and later it is an independent subject in the curriculum (e.g. the situation of the geography depends on the type of the upper secondary school). It is an independent subject in Austria, Slovenia, Hungary, Ireland, the Netherlands, Finland, Germany and Great Britain. In some countries, geographical knowledge is part of social sciences (Germany, Sweden) or natural sciences (Finland), or both (e.g., Slovenia, Great Britain, Austria, Ireland, the Netherlands, Hungary). Therefore often not the specialized teachers educate the geographical knowledge and they pay less attention to geography. In some countries geographic knowledge is taught with other topics at some educational levels (Curić et al. 2007), e.g.

- In Austria with economics in the lower and upper secondary level,
- In France and Ireland with history at the upper secondary level,
- In Finland with biology at the lower secondary level,
- In Sweden it is included in the social sciences like history, religion and society.

### *4.3.2 Cartographical Themes in the Curriculum*

*Cartography* is "the art, science and technology of making maps, together with their study as scientific documents and works of art" (Neumann 1997, p. 29). *Cartography in schools*, as a section of cartography, deals with making maps or map-like representation for school purpose and forming the criteria and the base of didactic-methodical use (Bollmann and Koch 2002). *School cartography* is map-making for school use (Neumann 1997) and includes not only the cartographical skills but also the use of maps and map reading and the topographic knowledge. The aim of topographic education is not collecting pieces of information, but developing ability and independent search and arranging knowledge. The topographic fundamentals, location, explanation of places, attaching geographic knowledge are also important.

**Table 4.2** Cartographical themes and topic in schools

| Cartographical themes | Cartographical topics |
|---|---|
| The map | Definition of the map, map types, globe, atlas, scales and scale bar, units |
| | History of cartography |
| Cartographical representation | Generalization, methods of representation, colours on the map |
| | Representation of relief, symbols, geographical names |
| | Projections, distortions |
| Orientation | Directions, compass, north and their determination, orientation of the map |
| | Determining the position |
| | Coordinate systems: geographic coordinate systems, km grid |
| Measurements on the map | Determining the amount and coordinates |
| | Measurement of distances, distance, areas and angles |
| | Use of maps and atlases (finder grid, telemetry) |
| | Gathering information from maps (earth grid, points of compass, GPS, compasses), creating topographic profiles |
| Remote sensing | Satellites, satellite images |

*European Schoolnet (EUN)*, a virtual learning network, has got 31 members of Ministries of Education in Europe since 1996. Its aim is to bring about cooperation, innovation and exchange in teaching and learning to its key stakeholders: Ministries of Education, schools, teachers and researchers. The main educational portals and the Ministries of Education of the member states can be found in it. The open and free educational contents are available from the Learning Resource Exchange (LRE) portal for schools from many different countries and providers, including 18 Ministries of Education. The basis for this comparison is this website and some school books.

If we focus on *cartographical topic in schools* of e.g. Austria and Hungary (Table 4.2), we can declare that:

- At the age of 8–10, the third to fourth grades in *primary school*, pupils learn about orientation in time and space for the first time.
- In *lower secondary education*, in the fifth to sixth grade, cartographical themes are taught with the help of globe, maps, atlases and pictures during the first 5–10 lessons. Students learn most important fundamental elements: such as map types, direction and distance, orientation, map symbols, relief, use of maps and atlases (finder grid, telemetry) and gathering information from maps (earth grid, points of compass, GPS, compasses). In all themes local complements are added. Topography plays a role in processing school work. According to valid curriculum its task is connecting topographic knowledge with geographic content.
- In the *upper secondary education*, the 9th grade, pupils mostly revise cartographical knowledge learnt at earlier stages, but mainly make use of map reading and analyses. They use maps, air and space photos in studying different themes. Requirements are map reading, comparative analysis of thematic maps and explanation of space photos, different geographical exercises (e.g. determining northern direction and height, orientation on maps, code and measurement of distance, zone and standard time). In the next years there are no cartographic

fundamentals taught at this level but orientation and use of maps in Geography lessons become important, expression of ideas on the basis of obtained knowledge, logical comparative analysis and identification of phenomena of nature on contour maps are important. As the amount of topographic events grows and time in school is little, extra lessons are needed in spare time.

The *final exam* at the end of upper secondary school is different in the member states. In some countries there is only a written exam (Bulgaria, Greece, Cyprus, Lithuania, Portugal and Finland). But in most countries, it has both written and oral parts. The written exam is often compiled and valued by an external authority. But the same (internal or external) committee organizes always both parts of the final exam, e.g. (Eurydice 2009):

- In the Netherlands, internal written and/or oral exam is set by the school, the external written exam is organized by external authority,
- In Belgium, in the Czech Republic, Slovakia and Iceland the written exam is set by the teachers within the school,
- An internal and an external exam is obligatory in Greece,
- In Portugal there is only an external final exam,
- In Austria the school supervisor sets the questions for the written exam, and the teachers do it for the oral exam.

In some countries, e.g. Austria there is no uniform final exam, every school decides its own requirements. Beside three compulsory subjects one obligatory subject is allowed. It can be Geography as well. Cartographical fundamentals are

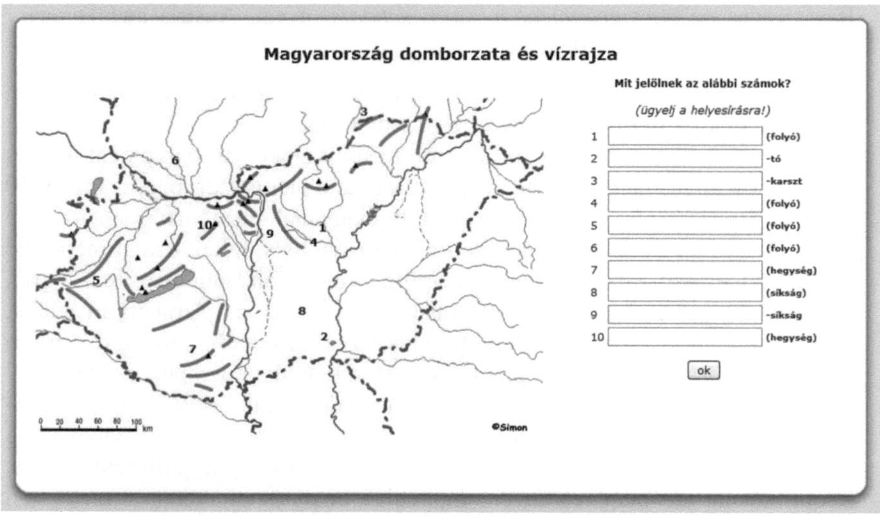


**Fig. 4.5** The *GeoLearn* is an interactive web application to practice spelling of geographical names. It corresponds to the Hungarian curriculum requirements for the written final exam (http://geolearn.fw.hu)

out of question. Only map reading and local knowledge are important. In other countries, e.g. in Hungary there is a new, uniform, two-tier system of secondary school final examination. Beside the original secondary school final examination, students can choose a raised level exam which serves as an entry exam to universities and colleges. One of the most popular optional subjects is geography. The written exam contains topographic knowledge (15% in medium level, 20% in raised level—Fig. 4.5) and exercises (determining place and height, scales, local and zone time). To answer test questions, school atlases can be used. Spoken exam also has cartographic and astronomical topics. The requirements of the exam are: the use of maps, analysis with the help of atlases. Topographic knowledge is based on a list issued by the Ministry of National Resources (OKM 2005).

## 4.4 New Media in Education

We can obtain this knowledge in many ways with the help of numberless educational tools. Schools use great number of traditional means. But nowadays we can choose from wide range of educational tools taking into consideration the opportunities of informatics. As internet and especially interactive web pages are new resources to help our daily lives. Since students today tend to surround themselves with modern technology, beside *simple classroom aids* (books, workbooks, visual aids: maps, atlases, globes, foils, slides, etc.) they may increasingly expect it in the classroom as well. The various digital sources play more and more important role in the education of geographic and cartographic knowledge.

The numbers of *CD-ROMs and DVD-ROMs* have increased which can be used successfully for teaching continents, countries and some topics (astronomy, meteorology and topography). The problem is that only a few suit the demand and content of geography teaching.

*Internet* is becoming more and more important. The web cartography can also distinguish different types of maps according to their complexity (Table 4.3). We can find plenty of geographical portals, databases or web pages containing tests, puzzles, quizzes and blank maps which can be used to prepare for the school lesson or tests (Fig. 4.6). The web cartography can have different roles and advantages in the geography and cartography education: simple, fast and cheap source in internet; it ensures a permanent presence all over the world; more effective communication with its multi-media, interactive and 3D elements; with the popular mapping service and entertainment components arouse pupil's attention; educational content in the form of tasks and games; you cannot only read, see, hear, but also try it; available both at home and at school.

With the new possibilities of the *web mapping 2.0*, method and content of teaching expands and a new quality in teaching and training comes into life. These websites can be founded on a concept of *edutainment* (education and entertainment), which is the method of attractive teaching and learning with multimedia applications. On the other hand, it can be used to enhance the

**Table 4.3** Web mapping software, techniques and standards (Plewe 2007; Haklay et al. 2008; Gartner 2009; Simonné-Dombóvári et. al. 2010)

| Generations of the web mapping | Examples for web technologies |
|---|---|
| *Statistic maps* e.g. Xerox PARC Map Viewer (1993), National Atlas of Canada (1994), Tiger Mapping Service (1995), MapQuest (1996) | HTML 1.0 (1993)<br>Java 1.0, Argus MapGuide, HTML 2.0 (1995) |
| *Maps with extended interactivity and functionality* e.g. TerraServer USA WMS for satellite and aerial photos (1998), Tirolatlas (2001) | Intergraph Geomedia Web 1.0 and Macromedia Flash 1.0 (1996)<br>ESRI ArcIMS 1.0, UMN MapServer 1.0 (1997) |
| *Web Mapping Services* e.g. Google Maps, housingmaps.com: 1. MashUp and Microsoft Virtual Earth (2005) | ESRI MapObjects Internet Map Server (1998)<br>Flash Player 5 (2000)<br>SVG 1.0 W3C Recommendation (2001) |
| *Virtual globes* e.g. NASA World Wind (2003), Google Earth (2005), Wikimapia and ESRI ArcGIS Explorer (2006), Google MyMaps and GoogleStreetView (2007) | Google Maps API 1.0 (2005)<br>Microsoft Silverlight, Apple iPhone (2007)<br>Google Earth Browser Plug-In, KML 2.2 W3C Standard (2008) |
| *Mobile web maps* e.g. Wikitude and Google Navigation (2009) | G1 mobile with GPS and compass (2008) |

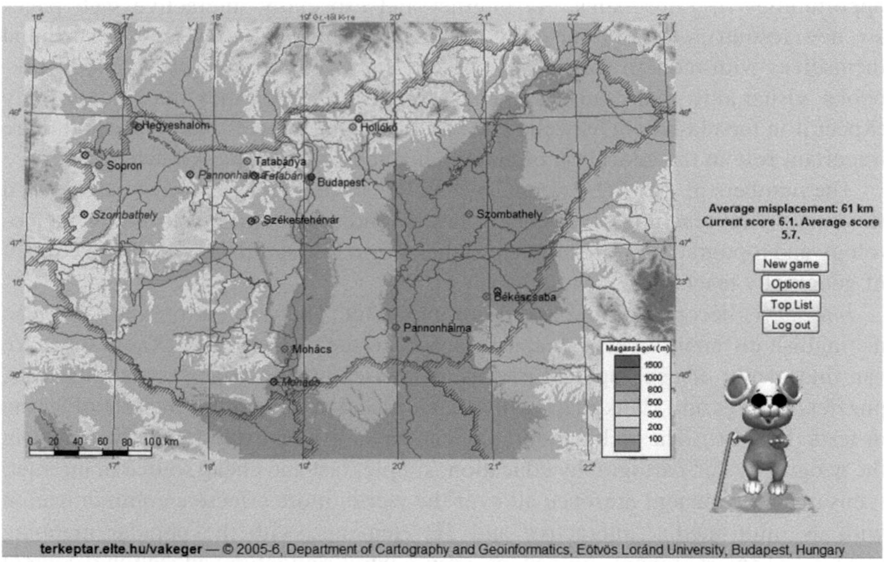


**Fig. 4.6** Result evaluation page—*The Blind Mouse*, mute map game for enriching the topographic knowledge of pupils (http://vakeger.elte.hu/)

attractiveness of cartography in different age groups. These can be used universally but with their help it is playful and stainless to memorize, deepen, test knowledge (Fig. 4.7).

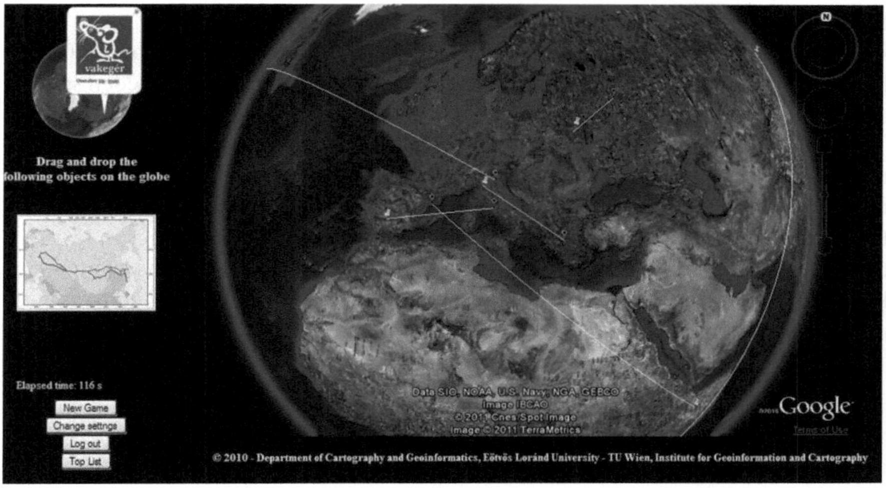


**Fig. 4.7** The game starts—*The three-dimensional version of the Blind Mouse*, mute map game with the Google Earth plug-in with the introduction of new functions of web mapping technology (http://vakeger.elte.hu/)

## 4.5 Conclusions

Countries with dissimilar past, development and tradition have joined the European Union and have come closer while regional differences are going to disappear. Their aim is the approach to uniform compulsory education, teaching competences, openness and individualization of learning. Unifying systems value international comparisons. Studying and comparing the everyday practice in these countries can help with finding better ways to teaching.

The first things that can be easily compared are the models how different levels of compulsory education follow one another, when they join, when they split. We can recognize three variations. As far as I see, for students too short first period or too many changes bring early selection, while a longer first period means “better foundation”, a solid knowledge.

The more number of lessons guarantee the more experience. But the role and the teaching time of geography has been reduced both in primary and secondary education in some of the member states. Nowadays the main aim of teaching is giving competitive and renewable knowledge. Its sense is put into the curriculum. If it is controlled, every school has the same possibility. If it is defined by schools, teachers of different schools offer different knowledge for students. They may cause principal differences.

Teaching cartography is one little piece of these systems and geography education. If Geography, Cartography is integrated, it suits lower levels of education. It may be not enough at higher level, when detailed knowledge of Geography is needed with the help of specified teachers. Teaching focuses on skills, preparing lifelong learning these days. In schools practice has greater value than paper knowledge. We can obtain this knowledge in many ways with the help of numberless educational tools. New techniques, active learning strategies are needed such as cartographical educational tools on the web.

With the new possibilities of the web mapping 2.0, method and content of teaching expands and a new quality in teaching and training comes into life. The role of new technologies and web cartography is becoming more and more important in our daily lives as well as in education. It brings new possibilities in learning and practicing in school or at home. It can help to teach one of the most important knowledge in Geography, to show the spatial and temporal trends of geographical phenomena and processes.

## References

Bollmann J, Koch W (2002) Lexikon der Kartograpie und Geomatik. Spektrum Akademischer Verlag, Heidelberg, pp 313–314

Curić Z, Vuk R, Jakovčić M (2007) Geography curricula for compulsory education in 11 European countries—comparative analysis. http://hrcak.srce.hr/file/39111. Accessed 29 Mar 2010

Eurydice (2002) Structures of education, vocational training and adult education systems in Europe. http://eacea.ec.europa.eu/ressources/eurydice/pdf/041DN/041_AT_EN.pdf. Accessed 8 Jan 2009

Eurydice (2009) Key data on education in Europe 2009. http://eacea.ec.europa.eu/education/eurydice/documents/key_data_series/105EN.pdf. Accessed 3 Feb 2011

Eurydice (2011) Information on education systems and policies in Europe. http://eacea.ec.europa.eu/education/eurydice/index_en.php. Accessed 3 Feb 2011

Gartner G (2009) Web mapping 2.0. In: Dodge M, Kitchin R (eds) Rethinking maps, Routledge studies in human geography. Routledge, London/New York, pp 68–82

Haklay M, Singleton A, Parker C (2008) Web mapping 2.0: the neogeography of the GeoWeb. Geogr Compass 2:2011–2039. doi:10.1111/j.1749-8198.2008.00167

Herodot (2007) Aspects of the state of geography in European higher education. http://www.herodot.net/state/state-geog-report.pdf. Accessed 10 Oct 2009

Kőfalvi T (2006) e-tanítás, Információs és kommunikációs technológiák felhasználása az oktatásban—Alapismeretek a tanári mesterségre készülők számára. Nemzeti Tankönyvkiadó, Budapest

Kozma T (1997) Például Ausztria. http://www.oki.hu/oldal.php?tipus=cikk&kod=1997-10-vt-Kozma-Peldaul#top. Accessed 11 Oct 2008

Lannert J, Mártonfi Gy (2003) Az oktatási rendszer és a tanulói továbbhaladás. http://www.oki.hu/oldal.php?tipus=cikk&kod=Jelentes2003-Tovabbhaladas. Accessed 11 Oct 2008

Mezei GY, Szebenyi P (1998) A közoktatás rendszere. OKKER Kiadó, Budapest

Neumann J (1997) Encyclopedic dictionary of cartography in 25 languages. K G Saur, München

OKI—Oktatáskutató és Fejlesztő Intézet (2000) Az oktatás és az európai integráció, http://www.oki.hu/oldal.php?tipus=kiadvany&kod=Oktatas. Accessed 19 Apr 2010

OKM—Oktatási és Kulturális Minisztérium (2003) 243/2003. (XII. 17.) Korm. rendelet a Nemzeti alaptanterv kiadásáról, bevezetéséről és alkalmazásáról. http://www.okm.gov.hu/letolt/kozokt/nat_070926.pdf. Accessed 8 Jan 2009

OKM—Oktatási és Kulturális Minisztérium (2005) Az érettségiről tanároknak. http://www.okm.gov.hu/letolt/kozokt/erettsegi2005/tanaroknak/foldrajz/foldrajzbe.htm. Accessed 8 Jan 2009

OKM—Oktatási és Kulturális Minisztérium (2009) Az Európai Unió és az oktatás, képzés. http://www.okm.gov.hu/main.php?folderID=861. Accessed 26 Jan 2010

OKM—Oktatási és Kulturális Minisztérium (2010) Az oktatás területét érintő európai uniós jogszabályok listája. http://www.okm.gov.hu/jogszabalyok/oktatas-teruletet. Accessed 10 Oct 2010

Ormándi J (2006) Összehasonlító pedagógia. APC-Stúdió, Gyula

Plewe B (2007) Web cartography in the United States. Cartogr Geogr Inf Sci 34:133–136

Simonné-Dombóvári E, Schmidt M, Gartner G (2010) Kartenanwendungen im Web. HMD—Praxis der Wirtschaftsinformatik 276:59–67

TIMSS (2007) Összefoglaló jelentés a 4. és 8. évfolyamos tanulók képességeiről matematikából és természettudományból. http://timss.hu/KMEO-TIMSS-2007.pdf. Accessed 10 Oct 2009

UNESCO—United Nations Educational, Scientific and Cultural Organization (1997) International Standard Classification of Education ISCED 1997. http://www.uis.unesco.org/TEMPLATE/pdf/isced/ISCED_A.pdf. Accessed 8 Jan 2009

Ütőné Visi J (2009) A földrajz tantárgy helyzetét és fejlesztési feladatait feltáró tanulmány—Kérdőíves elemzése felmérés alapján. http://ofi.hu/tudastar/tanitas-tanulas/foldrajz-tantargy. Accessed 8 Jan 2009

# Chapter 5
# Cartography at Elementary School Level: Continuing Education of Teachers and Experiences in the Classroom

**Maria Isabel Castreghini de Freitas**

**Abstract** The main aim of this article is to present a methodology for the continuing education of teachers in the area of Cartography, based on the formation of study groups arranged by tutors at the Center of Continuing Education for Mathematics, Science and Environment at São Paulo State University (CECEMCA/UNESP). The paper also aims to share some information about the Cartography activities related to the preparation of sketch maps, led by a teacher who attended the Continuing Education Course. In the activity related to the sketch maps of the way to school, three pieces of work were selected to illustrate the diversity of representation by the pupils, in accordance with their previous experiences and their level of cognitive maturity. The results of the study highlighted the fact that teachers, in general, find it difficult to prepare records, which limits their effectiveness and hinders the transformation of the practical work in the classroom.

## 5.1 Introduction

Today, Cartography is a mandatory and widely taught discipline in Geography degree courses and aims to provide graduates and undergraduates with knowledge that will allow them to read and interpret cartographic, systemic and thematic documents in analogical or digital form.

Whilst observing the work of professionals at elementary school level, we noticed that there was an enormous discrepancy between the material that was presented in their classes and a large amount of material from the textbooks, which, as a rule, contain cartographic material related to Geography. We have observed,

M.I.C. de Freitas (✉)
Department of Territorial Planning and Geoprocessing, IGCE/UNESP—Universidade Estadual Paulista, Rio Claro, Brazil
e-mail: ifreitas@rc.unesp.br

L. Zentai and J.R. Nunez (eds.), *Maps for the Future*,
Lecture Notes in Geoinformation and Cartography 5,
DOI 10.1007/978-3-642-19522-8_5, © Springer-Verlag Berlin Heidelberg 2012

through experience of Cartography courses taught throughout Brazil, that a significant number of Elementary School teachers are reluctant to give their pupils cartographic activities and exercises. They claim that this is due to limited education in this area. This should be taken into consideration given that the majority of these teachers come from Pedagogy courses, which do not have a tradition of offering subjects that deal with these topics.

Even for the Geography graduate, working in the Cartography area it is not an easy task. Despite the significant number of hours dedicated to Teaching Practice, courses such as School Cartography are rarely present among the compulsory disciplines that make up the core curriculum of teacher training courses in Brazil. We consider that the further education of graduates should take into account the specialties of school education, which, according to Santos and Kulaif (2006), should be one of the instruments that allow a human being to become a conscientious and free individual. For the author, well-taught School Geography teaches a pupil to become conscious of his/her spatial reality, for which the intervention of the teacher is necessary in terms of integrating local spatial logic with global spatial logic in his/her classes.

In the attempt to give credit to such a thought, Cartography and other related disciplines are fundamental in educating the graduate. The authors stress that *cartography being a visual medium (language) is a powerful tool in creating a school geography that is more dynamic at elementary and secondary school level* (Santos and Kulaif 2006).

## 5.2 Objectives

The main objective of this article is to present a methodology of continuing education for teachers from CECEMCA/UNESP, in the area of Cartography, based on the formation of study groups conducted by tutors. It also aims to share some of the activities related to the preparation of sketch maps, conducted by a teacher who attended the Continuing Education Course for Primary School Teachers. This study aims to illustrate the possibilities of representation the pupils came up with in the early years of elementary school as results of the experiments in the classroom organized and applied by teachers who participated in the study groups. These participants can contribute as specialists in the area and in Primary School teachers.

## 5.3 Continuing Education Course for Teachers and the Practice of Recording

In the book, Teacher Knowledge and Professional Education by Maurice Tardif (Tardif 2010) the author asserts that it is not possible to disassociate teacher knowledge from aspects involved in their daily work routine, from the teacher's personal life and professional background:

> In reality, in the area of employment and professions, I don't believe that we can speak of knowledge without relating it to the key factors and the work context: knowledge is always the knowledge of someone that works on something with the purpose of achieving some goal (objective). Furthermore, knowledge is not something that floats in space: a teacher's knowledge is his/her knowledge and it is related to his/her person and to his/her identity, to his/her life experiences and professional background, to his/her relationship with the students in the classroom and to other individuals in the school, etc. Therefore, it is necessary to study it by relating it to these essential elements of a teacher's job.

According to the author, teacher knowledge is social knowledge because it consists of shared knowledge among a group of agents, their peers, with a similar level of education initially, that operate in the same organization and are subject to similar work situations, such as the programs, the subjects and the rules of the teaching institution.

Taking into consideration the views of the author and our own experience, we can consider the way for us, to have success in teaching practices, which include, besides the initial stages, the continued education of teachers exercising their profession, is to take a closer look at the subjects taught and the teaching practices used:

> What is needed is not just the dispensation of disciplinary logic from teacher training programs, but at least open a larger space for professional education logic (initial and continuing), which recognizes the students (and the teachers in continuing education) as subjects who have knowledge and not simply as virgin spirits to which we simply provide disciplinary knowledge and procedural information without doing a thorough job related to cognitive, social and emotional beliefs and expectations, through which future teachers (and teachers in continuing education) receive and process this knowledge and information. This professional logic should be based on the analysis of practices, tasks and knowledge of professional teachers, and it should be a process of reflective focus, taking into account the actual teaching conditions and the strategies used to eliminate these factors in operation.

The structure of the proposal for the continuing education courses conducted on the premises of CECEMCA/UNESP came closer to the one presented by Tardif (2010). They are based on the interaction of knowledge of teacher through practices, and exchanging experiences in study groups. The main purpose of study groups is to interact with teaching practices, considering the cognitive aspects, the reality of life and work, not only in their school context.

The teacher in the classroom, as a rule, works alone. *In this setting, he/she never gets an immediate response. The silent understanding that accompanies this makes it difficult to know how it was, how well his/her discourse is being received* [...] (Fontana 2003).

The exchanges and interaction in the classroom are critical and need to be encouraged as they are during continuing education. The results of these exchanges should be recorded in the form of texts or narratives that detail not only the events involving the teachers and their professional performance, but also the questions, struggles and reflections about the teaching practice.

The act of recording makes high demands on the teacher, because it differs from the daily routine in the classroom by allowing our interaction with the act of teaching, recording our feelings arising from it, recording what remained in our memory, which stimulates and incommodes in the quest for transformation and transcendence:

> In the written report, we document our work of interpretation, through its gestures and words, the evidence of the relationship they share with education, whether it is to evaluate it or us, or to scrutinize our teaching practices (Fontana 2003).
>
> [. . .] None of us was prepared and ready to do it and nor was it necessary, as they made (and make) us believe. When we were sharing experiences, it became evident that the record as well as all the activities that make up the list of things that our teachers have to do, go beyond the demands in the course of our professional training, producing complex and arduous activities, which include difficulties that need be learned and developed in their principles and fundamentals (they are those that guide the procedures) (Fontana 2003).

In detailing some aspects of the record, the author stresses that there are many ways that you can get these records, asking questions, asking our peers for help, sharing this knowledge with our colleagues, provided that pupils, together with a teacher, dispense with routine practices whose meaning we had forgotten (Fontana 2003).

Given the statements exposed here, the record is a fundamental tool for the teacher, giving concreteness to the ideas, fears, reflections, certainties and uncertainties arising from the practice in the classroom. It should be seen as an ally of the teacher who seeks self-knowledge and understanding of his/her profession, striving for continuous improvement of his/her performance in the education of children and youth.

Training for Primary School teachers is not restricted to their initial training and continuing education courses. As stated by the researcher Maria Lucia Giovani (Giovani 1998) in a study on the formation (education) of teachers and on the role of the university in this process:

> It is recognized that the training of teachers and education specialists cannot be achieved through accumulating information, attending courses and acquiring techniques, rather by learning and real-life practice, individual and collective, critical reflection on practices and work contexts, providing opportunities for reconstructing professional and personal identity. It also recognizes the importance of "knowledge through experience" and providing opportunities to exchange experiences or "knowledge sharing" as a starting point for a new professionalism of staff in service. It is, above all a starting point, to initiate and sustain the effort of active appropriation of theoretical knowledge that supports and guides the competency to be put into practice (Giovani 1998).

Thus, learning on the job, *knowledge through experience*, whether it be individual or collective, should be viewed as opportunities for the stimulation and the appropriation of theoretical knowledge and instruments that enable the transformation of teacher practices in the classroom.

Also Freire (2005) in his lengthy debate of *right thinking* emphasizes the importance of thinking and doing together, sharing, communication between teacher and pupil, a *communicative act*, whose understanding is *co-participated*:

> Right thinking is not—one isolating themselves, one hiding himself away in a secluded place—rather it is a communicative act. There is no right thinking without understanding and this understanding, from a correct thinking point of view, is not something transferred but co-participated. [. . .] The task facing the individual, who thinks right, is not

transferring, depositing, offering and donating to another, to become a victim of his own thinking, the intelligibility of things, facts and concepts. The task of the educator, who consistently thinks right, is employing, as a human being, the irrecusable practice of understanding reality via objective logic, challenging the student with whom he/she communicates and producing understanding of what is being communicated. There is no intelligibility that is not communication and intercommunication, and that is not based on dialogicality. Right thinking is dialogical and therefore not controversial (Freire 2005).

The authors presented here agree about knowledge through practice, gaining knowledge through means of challenge and communication, whose experiences can and should be shared through records of teaching practices. According to Freitas and Yokoro (2009), in the article Cartography in Continuing Education of Teachers: Myths, Fears and Living Experiences, which presents some of the results obtained in training activities of teachers from CECEMCA/UNESP, the teacher's training will only be of an effective, definitive and permanent character when they are aware that they are first and foremost responsible for their education and retainers of the knowledge and experiences arising from their teaching practices, which must be shared and improved for the consolidation of knowledge in their pupils.

## 5.4 Experiences with School Cartography

Cartography is presented in Geography as a discipline whose theoretical and technical foundations allow for the exploitation and enhancement of the capabilities of primary school pupils, such as their observation skills, acquisition of knowledge, power of explanation, comparison and representation of space and its transformation, by means of elementary topological relations, reading and interpretation and preparation of sketches and maps to represent characteristics of their living space, personal space, as well as different landscapes and geographical space.

Lívia de Oliveira's pioneering work, which lays the foundation for School Cartography in Brazil was based on the theoretical framework of Piaget and was the reference point for a significant amount of the work that researchers produced afterwards.

In her thesis for full professorship, defended in 1978, Oliveira presents the Methodological and cognitive framework for maps (Oliveira 1978), a contribution to laying the methodological foundations of map study in Geography:

In this study we set out to approach the map from a cognitive and methodological point of view. It was designed, therefore, in order to contribute to the foundations of a map methodology (Oliveira 1978).

The author adheres to Piaget's studies on the child's intellectual development over space, *which states that topological spatial relations are the first to be established by the child, both on a perceptive and representative level, and it is in the framework of topological relations that the Projective and Euclidean relations will be developed* (Oliveira 2008).

From the perspective of education, it is considered *that a qualitative analysis of the map is justified from a cognitive point of view*, requiring the teacher to have *flexible teaching guidelines that are easy to manage and of low cost* (Oliveira 2008).

We agree with the assertions of Oliveira (2008), when considering that *the value of the map is determined by what the teacher intends to do with it,* meaning that it is up to the teacher, who possesses knowledge of this instrument, to use and apply this *model of reality* to situations that arise during their lessons (Oliveira 2008).

According to Almeida (2003):

> Piaget, with the support of a team of researchers conducted several studies that enabled him to create one of the more complete genetic theories on the cognitive development of humans. Even though today, in the light of other theories, Piaget's proposal suffers certain restrictions, his studies remain fundamental in terms of their representation of space.

In a study of Geography for a child's learning, Callai (2005) points out the importance of cartography for the education of children at Elementary School:

> One way to read space is through maps, which are the cartographic representation of a given space. Scholars of cartography teaching/learning consider that, for an individual to be able to read space critically, it is necessary for him/her to know how to interpret the actual/concrete space and as well to be capable of interpreting of his/her representation, the map. It is, of course, commonly understood that the individual who knows how to make a map will be in a much better position of being able to read a map. Drawing routes, paths, plans of the classroom, of their home, of their schoolyard may well be the beginning of the student's work for ways of representing space. These are tasks that children generally perform in their early years at school, but it is worth remembering that the interesting thing is that they base them on concrete and real data/facts and not on things that are imagined. That is to say, that they are trying to represent that which actually exists.

The author emphasizes here that which was pointed out by Oliveira (2008): the need to construct models that are close to the reality of the child, their concrete spatial domain, representing through sketch maps, plans and models their every-day living environment.

In this study, with reference to the cited authors, we consider it necessary that cartographic activities be experienced and incorporated into the everyday practices of teachers of the early years at Elementary School. They should be prepared for this when doing their undergraduate courses, and courses undertaken once in the profession, called continuing education courses. Thus, the disciplines related to topics such as School Cartography and Teaching Methodology for Cartography, in undergraduate courses, should be valued more, whether they be Geography or Pedagogy to ensure that children and youth are adequately instructed in the subject.

## 5.5 Methodological Guidelines

In order to develop our work, we undertook a literature review, consulting documents related to the initial and continuing training of Primary School teachers in the areas of Education and Cartography.

The basic data for the research were drawn from our experience in creating and coordinating the CECEMCA/UNESP continuing education courses on the theme Cartography and the Environment, especially the reports prepared for UNESP and MEC from 2007 to 2009 as well as the reports of the educators, tutors and professors involved in the process founded in UNESP (2008, 2009). We still had access to final reports of the participating teachers, who developed practical activities with their pupils in the classroom, some of which were selected to give concreteness to the ideas expressed here.

To develop the research we referred to the qualitative methodology, which is explained in Bogdan and Biklen (1994):

> We use the term qualitative investigation as a generic term that includes numerous research strategies that share certain characteristics. The data collected are called qualitative, which means rich in descriptive de-tails regarding people, places and conversations, and complex in terms of statistical analysis. The issues to be investigated are not established through operationalizing variables; rather, they are formulated with the objective to investigate the phenomena in all their complexity and natural context. Although the individuals who carry out qualitative research might select specific issues as they collect the data, the approach to the research is not designed to answer previous questions or test hypotheses. They favor, in essence, the understanding of behaviors based on the point of view of the subjects of the investigation.

Thus, stories, photos, drawings and descriptions of tasks performed by the teacher and his/her pupils, are presented in the attempt to comprehend how effective the education of the teachers in Cartography and the Environment was and understand the results of the practices developed throughout the course.

The Center for Continuing Education in Mathematic, Scientific and Environmental Education (CECEMCA/UNESP) is a UNESP Extension Program run by the Dean's Office, which began its operation following approval of the project submitted by UNESP to the Ministry of Education (MEC), which culminated with the creation of CECEMCA/UNESP in March 2004. From this date on, CECEMCA went on to set up the National Network for Continuing Education of Teachers of the Secretariat of Elementary Education (SEB) of MEC.

All the teaching material prepared by the CECEMCA team makes up the SEB/MEC network file. In this experiment of educating teachers about Cartography and the Environment, through the continuing education courses at CECEMCA/UNESP, we referred to the book Cartography and the Environment by Freitas (2005).

There was a condition in these courses that all registered teachers had to be part of a study group, in which there was a mediator appointed by the Secretary of Education (named as tutor), who had prior training in a specific course. We placed the responsibility on the teaching leaders of the Municipality, the Secretary of Education and his team, to choose the local tutor. The tutor was responsible for preparing the learning activities to act as coordinator of the Study Group, composed of 10–30 teachers of his city and region.

The training courses for tutors ranged from 40 to 120 h, while the teacher training courses mediated by tutors were from 40 to 180 h, depending on the demands of the systems of teaching partners. One standout feature of the training courses for the tutors and teachers was the semi face-to-face format, with part of the work carried out in long distance mode. In the case of the tutor-training course, we adopted a The Long Distance Education Environment (TelEduc) to facilitate communication between the educators at CECEMCA and the teachers-tutors. Tel-Educ Environment is a Virtual Learning Environment (VLE) created by Centro de Computação (Center of Computation) of UNICAMP (University of Campinas) in 1997, with various tools for virtual interaction such as a menu of activities and tasks, posting of reading material, slides, films, chat room, e-mail, agenda, place for posting tasks among others, that permit the creation and administration of long distance courses, partially face-to-face and also used as support in the face-to-face classes.

In their continuing education courses, the tutors were prepared to coordinate study groups for teachers, and were also responsible for the descriptive activities in the Cartography book, organization of practical activities, debates and discussions, monitoring the group with the central themes: Cartographic Representation of the Environment, Thematic Cartography in Environmental Studies, Remote Sensing in Environmental Studies and Environmental Awareness and Educational Excursions.

The local tutors were instructed to draw up an agenda for each meeting to be held, which was then sent to the center through the Virtual Environment (Tel-Educ) before each meeting. Secondly, the tutors would post up personal details regarding the meetings, such as the activities undertaken and records (written reports) of the participating teachers on a board.

The written reports made at the end of each learning activity, face-to-face or in distance learning mode, assisted part of the discussions in the study groups and made it possible for local tutors under the supervision of the trainer and coordinator of the course, to modify the activities at each meeting, according to the needs and concerns reported.

## 5.6 Reports of Teachers and Educators

This experience of integrated education can open new possibilities for the performance of teachers in the classroom. In the stories that follow appear the mechanisms that the teachers adopted to carry out their work, reflecting on the teaching material available and the necessity to adapt it to the needs of their pupils with respect to age and their life context. Given the ready acceptance of the course by teachers, it was interesting to note the initial comments of some teachers about the versatility of the material in the disciplines, because the study groups were comprised of teachers of different subjects and year levels. All names of teachers and pupils are fictitious to preserve their identities:

> I noticed the enthusiasm of teachers starting their study of cartography, the associations made between theory and practice made them reflect and understand the great need to develop activities for Cartography, not only in Geography but in all areas of knowledge. The teachers thanked us for the opportunity to participate in the course offered, because in their reports they were happy to say that: “It was very valuable because up to this point my

> greatest resource was the textbook". Those words made me think and see that the course offered will be – as it was for me – a great new interdisciplinary development applied in context. Another participant concluded: "Knowing how to read, interpret, analyze and find one's place in the representation of the Earth, makes the student a human being, capable of guiding him/herself and defining his/her place (space)..." (Instructor Antonio, Naviraí-MS, 2007)

The difficulty of access to communication equipment via the Internet slowed down the training courses and resulted in many tutors delivering their reports late. The difficulty of reporting their experiences during the study group meetings, in either the face-to-face or long distance mode meetings, was shared with the instructors from CECEMCA online.

Another differentiating aspect of the training courses concerns the preparation of the reports by the tutors, and we would like to include some remarks by an educator from CECEMCA:

> Sometimes they were actually reports, they mentioned that they had done a certain activity in the classroom and (that) sometimes the class had difficulty under-standing (the content)... These results came to us in two ways: reports from the face-to-face meetings (once a month) and through TelEduc (every week). Thus, most of them were delivered in person because of the difficulty they had registering and sending them via TelEduc, which was a little vague. When we went to the face-to-face meetings it was emphasized even more, with more detail and so (we realized that) not only the students were having difficulties, but sometimes the instructors—when they were trying to teach the students—realized that there was something they had not understood. (Educator Rafael, personal information, 2010)

The activities developed in the study groups on Thematic Cartography and the Environment led to the proposal to conduct studies and experiments involving the group of teachers and their practices in the classroom. To illustrate these activities we selected one of the practices involving the preparation of a Sketch Map of the Route from Home to School, which made up part of an account by one of the teachers participating in the Continuing Education Course at CECEMCA/UNESP.

## 5.7 Sketch Maps of the Route from Home to School

One activity present in the practices conducted by teachers in training was the preparation by the pupils themselves of sketch maps and drawings of a space closely, usually a classroom, but also of other environments such as home, the home street, the route from home to school, among others.

In Piaget's theoretical systematizations, knowing means organizing, structuring and explaining reality based on our own experiences. Knowledge is always the product of the subject's action on the object. Therefore, the operation is the essence of knowledge: the interiorized action modifies the object of knowledge, imposing order on space and time.

> [...] Upon analyzing its principles it is possible to infer that the Piagetian thread of argument, that is, the basic orientation of his work, expresses itself in the idea that knowledge does not originate in perception, but in action. (Palangana 2001)

From the activities related to sketch maps of the route from home to school carried out with pupils of fourth year, we selected three pieces of work that illustrate the diversity of possibilities of representation by the pupils according to their previous experiences and their stage of cognitive maturity to represent a mind map showing the home-school route on a sheet of paper.

Figures 5.1, 5.2 and 5.3 correspond to the representations of the route from home to school prepared by pupils of a teacher called Agatha (Piracicaba-SP), who de-scribes the activity undertaken in his report to the study group:

> In order that they would have a notion of a sketch map, I asked them to draw the route that they take from their home to the school, including the important places along the route so that I could get there. This activity had greater results with many details. (Teacher Agatha, Piracicaba-SP, 2007)

In Fig. 5.1, we see a drawing by a pupil called Gio, with a bird's eye view of the street and images of the house fronts facing upwards in order to represent clearly the objects along the route, where the most important thing is not the accuracy of the route itself, rather the depiction of the references showing where it is, such as the pub, the vegetable garden and the market, distinguished from the common houses by the labels on their fronts and a sign (in the case of the vegetable garden).

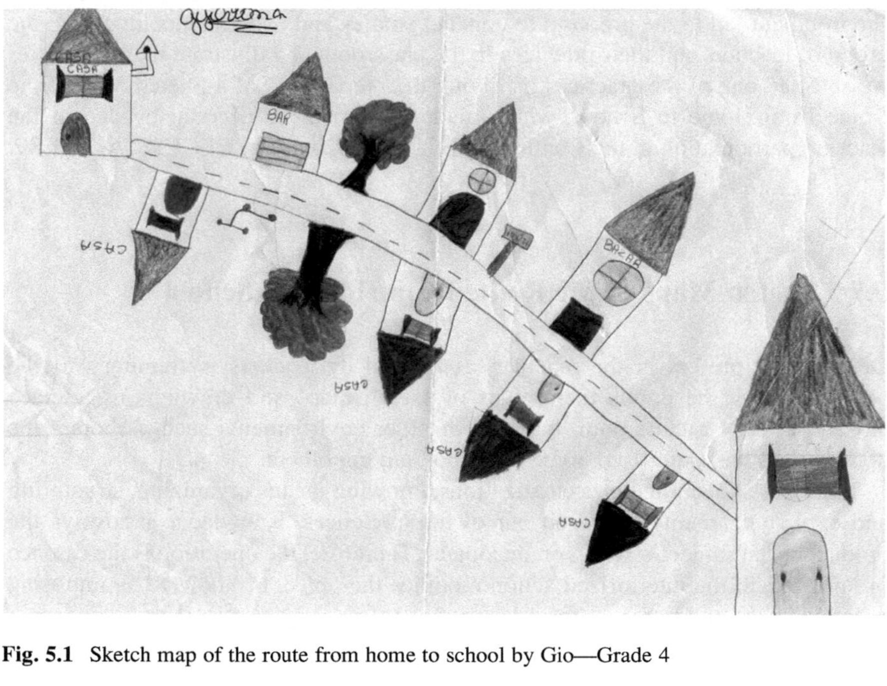


**Fig. 5.1** Sketch map of the route from home to school by Gio—Grade 4

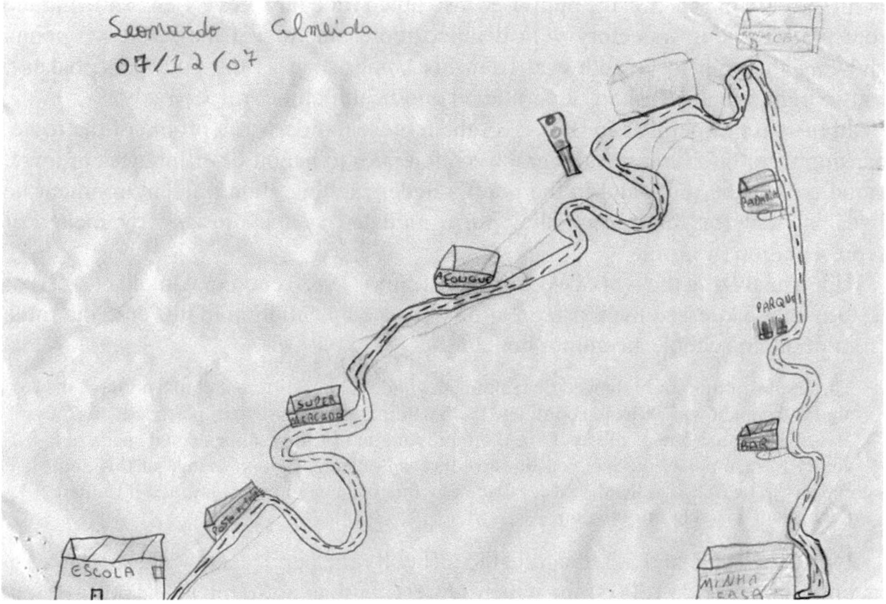


Fig. 5.2 Sketch map of the route from home to school by Leo—Grade 4

Fig. 5.3 Sketch map of the route from home to school by João—Grade 4

In the case of Fig. 5.2, the pupil Leo also opted for a bird's eye view, showing the route to be a long trajectory with detailed representation of the reference points showing the path to be followed from his home, such as the pub, park, bakery, traffic light, butcher's shop, supermarket and health clinic, respectively.

In the case of pupil João (Fig. 5.3), the representation was a profile of the route, causing us to infer that the topographic references to height or differences of level, stood out on the route and in the pupil's understanding about the environment he lives in, and for this reason they were included in the drawing by means of representation in profile.

Upon analyzing the work developed by Telmo (1986) cited in Almeida (2003) of a study that addressed the representation of space by children, in this case the route from home to school, the author concludes:

> The results of this study showed that, although children have different points of view, they are far from knowing how to coordinate them within a single system of perspectives. They also showed that one of the keys for the representation of three-dimensional space is the ability to manipulate the inclined lines to draw an object. The appearance of this ability appears to be linked to the discovery that an inclined plane represents implicit information rather than concrete information.

Figures 5.1, 5.2 and 5.3 illustrate the difficulty the pupils have with coordination as cited in Almeida (2003), in which representations are depicted from different points of view in the same representation.

In Fig. 5.1, the object "street" was represented as an aerial view, while the homes and businesses were facing up. In our view, the effect of this depiction of the buildings adopted by pupil sought to explain to the reader the differences between the objects that are on the route, e.g. the homes, the businesses and the vegetable garden, the latter represented by a clear sign. The pupil intuitively sought to explain in this representation, implicit information vertically that might not be visible in a representation from above. This fact confirms their difficulty in coordinating the representation of all the objects in a single perspective, which is due to them not having sufficient conceptual elements to present the representation vertically when preparing the sketch map.

In Fig. 5.2 we observed that through the drawing the pupil conveys to the reader the long trajectory he takes to get to school. He selects some common day-to-day objects, e.g. businesses and services (pub, bakery, butcher's shop, supermarket and health clinic) and uses symbols to make certain objects stand out on the route, such as a traffic light and an amusement park (represented by drawing two swings for children). His effort, in practicing cartographic generalization and simplifying the route, while informing the reader about the complexity of the path (represented by a series of curves) and its main objects for reference to assist movement and orientation in the space, makes his representation similar to the representation used in some touristic maps.

Figure 5.3 shows the relief as determinant factor for representation of the route from house to school, whose trajectory traverses a valley that separates the two main objects of interest. The representation of the road from a bird's eye view (the division of the two lanes indicated using a dashed line as the symbol) and representation of the vehicles in profile (cars and motorcycle), indicating horizontal or lateral vision as well as the representation of the houses in perspective, demonstrate the confusion pointed out by authors Telmo (1986) and Almeida (2003). The pupil's drawing is the only one that includes an allegorical drawing of the sun, sky and clouds, embellishing the representation and functioning as a backdrop for the image.

Analyzing the projects that were presented allows us to consider how, in the same class with pupils of a similar age (between 9 and 10), representations of the route from home to school can be presented in so many ways. The teacher should be aware of these situations, which are routine in the classroom, and value the way that each child tackles the task to prepare a sketch map, individually. The way of understanding and exploring our living space varies from person to person, according to their previous experiences. Referring back to Tardif (2010) mentioned at the beginning of this article, the knowledge of individuals, whether they are teachers or pupils, is directly linked to their identity and life experience.

Examples of sketch maps presented by children from Grade 4, show that the same proposed activity can be tackled and done differently, depending on how the pupils interpret the teacher's request, according to their cognitive development, their experiences and relationships with each other and with the environment in which they live. This diversity is repeated in many of the activities proposed in the training course at CECEMCA and reflect the diversity of people and the creative ability of each to face up to the daily challenges in the classroom, whether they are teachers or pupils.

Next we highlight a brief report by Aghata (classroom teacher), to which the previously presented Figs. 5.2 and 5.3 were annexed. Upon selecting the drawings presented and considering them "great results", led us to believe that the diversity of approaches taken by the pupils of the class to the proposal to prepare the sketch map was understood. It is not possible, however, to gauge the teacher's level of comprehension of the results achieved by their pupils in performing the activity based on the record made.

This kind of situation occurred several times in the continuing education courses offered by CECEMCA, indicating the teacher's difficulty to conduct an analysis of the effectiveness of their own work in the classroom. Being aware of the value of the material prepared by the pupils in question, allowed the teacher to select them to present to the study group, although it was not possible for her to interpret and analyze the results achieved from the practice in her record, leaving a gap between what was achieved and understood and documented in the course and the practice in the classroom.

## 5.8 Conclusions

The working environment of the continuing education courses at CECEMCA through the establishment of study groups, allowed those who accepted the challenge of this experience to grow as professionals. The compiling of written reports by the tutor and the teacher, although adopted in all teacher-training courses offered by CECEMCA, presented difficulties among those involved. Teachers, in general, have a greater mastery of speech than of writing, when it comes to giving expression to their reflections of the teaching profession and their experiences in the classroom. This leads us to believe that this kind of difficulty in preparing records limits teacher performance by putting obstacles in their way that they, individually and collectively, need to overcome to transform their practices in the classroom. These are some considerations based on the experiences of the activities undertaken in the continuing education courses conducted by CECEMCA/UNESP.

Children from the first stage of Elementary Education develop the activities in different manners, depending on how they interpret the teacher's request and according to their cognitive development and their previous experiences. The representations provided by the teachers were mostly displayed vertically with a flat or upturned view of the objects (e.g. houses and buildings), or the pictorial profile (horizontal view of the street with allegorical elements such as sun, clouds, among others), as illustrated in the examples presented previously. Drawings that depict the images from an exclusively vertical point of view are rare in this stage of cognitive development. The confusion with regard to the presentation of different objects is crucial in representation by this age group, which is consistent with the literature devoted to the theme.

Our expectation is that the experience reported here, from Brazilian elementary school continuing education courses of CECEMCA/UNESP and there resulting reflections can contribute with other countries with similar contexts.

## References

Almeida RD (2003) Do Desenho ao Mapa: Iniciação cartográfica na escola. 2nd edn. Contexto, São Paulo

Bogdan R, Biklen S (1994) Investigação Qualitativa em Educação: Uma introdução à teoria e aos métodos. Porto Editora LDA, Porto

Callai HC (2005) Aprendendo a Ler o Mundo: A Geografia nos Anos Iniciais do Ensino Fundamental, vol 25, n. 66, 227–247. Cadernos Cedes, Campinas. http://www.cedes.unicamp.br. Accessed 10 Dec 2010

Fontana RAC (2003) Como nos tornamos Professoras? 2nd edn. Autêntica, Belo Horizonte

Freire P (2005) Pedagogia da Autonomia: Saberes necessários à prática educative, 31st edn. Paz e Terra S/A, São Paulo

Freitas MIC (2005) Cartografia e Meio Ambiente, 1st edn. CECEMCA/UNESP/MEC, Bauru

Freitas MIC, Yokoro CMA (2009) Cartografia na Formação Continuada de Professores: Mitos, Medos e Experiências Vividas. Proceedings of 12 Encontro de Geógrafos da América Latina, vol 1, 1–12. Universidad de la República, Montevideo. http://egal2009.easyplanners .info/buscar.php. Accessed 03 Dec 2010

Giovani LM (1998) Do professor informante ao professor parceiro: Reflexões sobre o papel da universidade para o desenvolvimento profissional de professores e as mudanças na escola, vol 19, n. 44. Cadernos Cedes, Campinas. http://www.scielo.br/scielo. Accessed 03 Dec 2010

Oliveira L (1978) Estudo Metodológico e Cognitivo do Mapa. Novalunar, São Paulo

Oliveira L (2008) Estudo Metodológico e Cognitivo do Mapa. In: Almeida RD (ed) Cartografia Escolar. Contexto, São Paulo, pp 15–42

Palangana IP (2001) Desenvolvimento e Aprendizagem em Piaget e Vygotsky: A relevância do social, 3rd edn. Summus, São Paulo

Santos CE, Kulaif Y (2006) O Ensino de Geocartografia nos Cursos de Formação de Professores de Geografia no Brasil: apontamentos e reflexões. In: Proceedings of Simpósio Ibérico do Ensino de Geologia, Universidade de Aveiro, pp 126–161

Tardif M (2010) Saberes Docentes e Formação Profissional, 10th edn. Vozes, Petrópolis

Universidade Estadual Paulista UNESP (2008) Pró-Reitoria de Extensão Universitária. Programa do Centro de Educação Continuada em Educação Matemática, Científica e Ambiental. Relatório de Atividades CECEMCA/UNESP: 2006–2007. PROEX, São Paulo

Universidade Estadual Paulista UNESP (2009) Pró-Reitoria de Extensão Universitária. Programa do Centro de Educação Continuada em Educação Matemática, Científica e Ambiental. Relatório de Atividades CECEMCA/UNESP: 2007–2008. PROEX, São Paulo

# Chapter 6
# Cartography in Textbooks Published Between 1824 and 2002 in Brazil

**Levon Boligian and Rosângela Doin de Almeida**

**Abstract** The main focus of enquiry on this research is the historical evolution of Cartography contents from 1824 until 2002 and its role in the establishment of an educational geographic culture in Brazilian schools. It has identified curricular alternations, permanence and transformations of these contents during the investigated period, taking as source materials the official syllabus programmes, but especially the compendia and Geography textbooks addressed to students in the first year of secondary school. Through this socio-historical view of the curriculum, it was possible to evidence important epistemological differences between the scientific geographical knowledge and the educational geographical knowledge.

## 6.1 Purposes for This Study

This research investigated the hole that Geography textbooks, published since the beginning of the nineteenth century, and the curricular contents contained therein have in the socio-historical process of construction and reconstruction of an educational geographical knowledge considering the Brazilian educational system.

To achieve this aim, the evolution of Cartography contents presented by authors of textbooks and by the Brazilian official programmes to students in the first year of secondary school was analyzed and, through a spreadsheet of subjects, notions and cartographic concepts, it was possible to notice the alternations, the ruptures and the curricular transformations occurred between 1824 and 2002.

L. Boligian (✉)
Universidade do Vale do, Ivai, Brazil
e-mail: boligian@hotmail.com

R. Doin de Almeida
Universidade Estadual Paulista (UNESP), Campus of Rio Claro, Rio Claro, Brazil
e-mail: rda.doin@gmail.com

L. Zentai and J.R. Nunez (eds.), *Maps for the Future*,
Lecture Notes in Geoinformation and Cartography 5,
DOI 10.1007/978-3-642-19522-8_6, © Springer-Verlag Berlin Heidelberg 2012

## 6.2 Analysis of the Socio-historical Evolution in School Cartography Contents

A research conducted by Callai (1999) with a sector of students and teachers of secondary education in Brazilian state schools, in the middle of the nineties, revealed that, for the majority of people interviewed, Geography seems to be extremely fragmented and naturalistic. In this sense, names of places, their localization and mainly their natural characteristics are studied, once "Physical Geography" is a field in this subject considered to be the most "scientific", because is more observable and objective than "Human Geography". Besides, according to the above-mentioned research, for the interviewees, the ideas of "map" and "Cartography" are inseparable from "Geography". For them, these concepts are practically synonyms. Martinelli (2000) also highlights this aspect in his thesis, writing that:

> When we talk about 'maps', immediately we associate them with 'Geography'. It is a cultural aspect. Maps, therefore, would represent Geography, what is geographic. They would be Geography themselves. [. . .] In this sense, we can verify that a map is a symbolic representation of Geography. [. . .]

We understand that this representation in the imagination of people is a socio-historical construction whose origin is at schools, in the Cartography contents worked by Geography teachers. It is a knowledge dictated by the curricular programmes and especially by the textbooks published and used in Brazil in the last two centuries. During this period, the cartographic contents have been worked, roughly, in the volumes regarding the first years of secondary education. And these contents are our main focus of analysis on this study.

## 6.3 The Selection of Documents and Method of Work

Information and data collection for this research in documentary sources were carried out based on the Library of Textbooks (BLD) collection, located at the Faculty of Education of Universidade de São Paulo. We complemented the survey of textbooks, asking the most important Brazilian publishers for an authorization to look up their private collections, in a way to have access to the major works published in Brazil during the specified period. We had permission to consult the private collections from the following publishers: Saraiva, FTD, Scipione, Ática and Abril Educação.

Besides consulting the publishers' collections, we had access to available documents at the Documentation Department and Memory of Colégio Pedro II (NUDOM), in Rio de Janeiro, where we consulted the minutes of the Teachers Council, Geography compendia and theses regarding the school since its establishment. Colégio Pedro II was the first institution of secondary education created in Brazil, in 1837, in the same traditional mold of French collèges and licèes. Until the end of the fifties, the teaching staff of Colégio Pedro II was entrusted by the State with the formulation of the syllabus for every subject taught in secondary schools in the whole country. Consequently, the access to the above-mentioned documents was essential for the development of this work.

In relation to the criteria for selecting the textbooks used as source of information for the research, we used as major parameters: the works of authors mentioned in the researches of Issler (1973), Colesanti (1984), Vechia and Lorenz (1998), Gasparello (2002, 2006) and Vechia (2007), which, according to those specialists, stood out in the Brazilian educational circle until the third quarter of the twentieth century; and, with relation to the most recent textbooks, published from the seventies on, we opted for the ones that, according to information provided by the person in charge of the collection in the visited publishers, stood out in terms of adoption, both in private and governmental purchases in the last four decades.

In the event of textbooks of a same author that were republished during a long period of time—works called "editorial success"—we selected editions for the first year of secondary education, with intervals of some years between them, intending to verify developments carried out by the author in the way that the contents of Cartography were worked. Actually, in several cases, there was not a possibility to perform a sequential analysis of the editions due to shortages of some of them in the consulted collections, even in the own collections of the publishers. This fact referred us to a matter already mentioned by Bittencourt (1993) related to a characteristic of textbooks. As they are products that are quickly consumed, almost in a disposable way, in accordance with the changes in the curricular and market context, these publications create a paradox. Even if the titles have a great circulation—that is since the beginning of the twentieth century—they are badly preserved by the society in general, including the very publishers which produce and commercialize them in the whole country, because, in accordance with we could observe, few of them invest in the organization and conservation of their own collection of works already published.

Regarding the consulted curricular programmes, our main sources were the following documents: the book "Programa de ensino da escola secundária brasileira: 1850–1951", under the authorship of Professors Ariclê Vechia e Karl Michael Lorenz, from Universidade Federal do Paraná, in which the old official programmes from Colégio Pedro II were consulted, all of them drawn up and published during the period of Empire in Brazil; the dissertation "O ensino de Geografia através do livro didático no período de 1890 a 1971", under the authorship of Professor Marlene Teresinha de Muno Colesanti, published under orientation of Professor Lívia de Oliveira, in 1984, in which we looked up for the programmes of curricular reforms occurred during the First Republic and the New Republic in Brazil; the documents "Guias Curriculares Propostos para as Matérias do Núcleo Comum do Ensino de 1o Grau", drawn up by the Centre of Human Resources and Educational Researches "Professor Laerte Ramos de Carvalho" (CERHUPE), published in 1975, and the "Proposta Curricular para o Ensino de Geografia/1o Grau", drawn up by the Coordination Office of Studies and Pedagogic Principles, first edition published in 1988, both institutions part of the Education Department of São Paulo State, where we consulted the curricular programmes of Social Science and Geography, for the second half of the seventies and for the eighties and nineties, respectively. As in this period, the Federal Government, through the Law of Directives and Bases of Education, No. 5692/71, decentralised the establishment of curricular proposals, we tried to get information about cartographic contents from these documents because we believe that São Paulo has become a national reference; and lastly, the document "Geografia", for the fifth to the eighth grade, or the third and fourth cycles of Primary Education from the "Parâmetros Curriculares Nacionais" (PCNs), established and published by the Ministry of Education, in 1998 (Brazilian Department of Elementary Education).

Based on the criteria of selection already mentioned, for this study we consulted 13 (thirteen) official curricular programmes and 37 (thirty-seven) teaching materials, among compendia and textbooks.

After selecting the materials, the next step comprised of recording the data and information extracted that would be the basis for our analysis. After trying some resources, like making photocopies of books, we concluded that the best way to store the pages of the books, in order to subsequently analyze them, would be by means of digital photography. Therefore, we photographed the pages of the official programmes and the minutes of the Teachers Council of Colégio Pedro II, as well as the compendia and textbooks that had some concepts (especially the ones of public use or model exercise) linked to Cartography. Altogether, we accumulated around 5,730 images in JPG format.

The criteria to choose the Cartographic contents present in the curricula and textbooks were based on the analysis of the Simielli (1993, 1996, 2008), Almeida et al. (1997), Almeida (2003) and Le Sann (2005) studies, which helped us to compile a set of notions, concepts and cartographic subjects. As far as we understand, these contents were historically established as Cartography contents to be taught in Geography classes in Brazil. The Tables 6.1, 6.2 and 6.3 bring the contents extracted from the above-mentioned studies.

Based on the contents indicated in the mentioned tables (Tables 6.1, 6.2, 6.3), we drew up a cluster of notions, concepts and subjects, divided in some large thematic groups in order to consider a universe of School Cartography contents developed in the Brazilian secondary education. In the Table 6.4 are presented the groups created with their respective contents.

Based on this last table, we started reading the official programmes and the pages of the selected textbooks, trying to identify the presence of the chosen cartographic contents.

**Table 6.1** Cartographic contents proposed by Maria Elena Simielli (1993)

| Shapes | Proportion |
|---|---|
| Symbols | Measurements |
| Legend | Scale |
| Oblique view | Grids/tables |
| Vertical view | Coordinates |
| Representation in the plane | Maps (interpretation) |
| Generalisation | Aerial photographs |
| Laterality | Scale model |
| Direction/orientation | Localisation |

**Table 6.2** Cartographic contents proposed by Almeida et al. (1997)

| Book 2 | |
|---|---|
| Cartographic subjects | Knowledge relating to spatial representation |
| Scale | Conservation of measurement |
| | Conservation of distance |
| | Proportional areas |
| Cartographic projections | Projection of sphere in the plane |
| | Notions of circumference |
| | Concept of angle |
| | Notions of parallelism and perpendicularity |
| | Parallels and meridians |
| Topography | Orthogonal projection |
| | Planning of geometric solids |
| | Notion of plane and curves |
| | Perpendicularity between straight lines and planes |
| | Landforms |
| | Contour line |
| Projection in the plane | Conservation of a point of view |
| | Orthogonal projection from a vertical point of view |
| | Symbolisation |
| **Book 3** | |
| Cartographic subjects | Knowledge relating to spatial representation |
| Localisation | Location references on the globe: west-east and north-south directions |
| Orientation | Location and geographical orientations: parallels and meridians |
| Coordinates | Angle measurement |
| Scale | Notion of proportion |
| | Metric system |
| | Conservation of distance and measurement |
| | Calculation of distance |
| Topographical representation | Contour line |
| | Topographical outline |
| | Orthogonal projection from a vertical point of view |
| | Classification and connection between classes |
| | Relation between numerical and spatial proportions |
| | Visual proportion and variable visual size |
| | Visual order and variable visual value |
| **Book 4** | |
| Cartographic subjects | Knowledge relating to spatial representation |
| Localisation | Geographical references on the globe |
| Orientation | Orientation in the world map |
| Coordinates | Geographical coordinates |
| | Division of angles |
| Topographical representation | Ratio and proportion |
| | Perpendicularity between straight lines and planes |
| | Contour line |

(continued)

**Table 6.2** (continued)

| Book 4 | |
|---|---|
| Cartographic subjects | Knowledge relating to spatial representation |
| Scale | Ratio and proportion |
| | Cartographic projections |
| History of cartography | Cartographic projections |
| Thematic representations | Ratio and proportion |
| | Classification and connection between classes |
| | Relation between numerical and spatial proportions |
| | Visual proportion and variable visual size |
| | Visual order |
| Application of cartographic knowledge | Interpretation of aerial photographs |
| | Topographical map |
| | Land relief outline |

**Table 6.3** Cartographic contents proposed by Janine Le Sann (2005)

| General plane | |
|---|---|
| Main axes | Cartographic notions |
| Representation | Shape of objects |
| | Free drawing |
| | Photographs |
| Comparison | Scale |
| | Relation-dimension |
| Proportion | Standard measurement |
| | Perceptual measurement |

## 6.4 Results of the Documentary Analysis

The analysis of the selected official curricular programmes and teaching materials allowed a "visualization" of a historical evolution of cartographic contents in the last two centuries in Brazilian schools.

We perceived that since 1824 the contents are becoming increasingly diversified over time, with new notions, concepts and subjects being added to the syllabus prescribed, both in the official curricular programmes and in the teaching materials (compendia and textbooks) of Geography. It is well known how compendia and textbooks, in general, deal with more contents of Cartography than the curricula themselves, which bring a more simplified programme, except for the PCNs (1998).

In a more specific interpretation, we can observe the existence of a sort of "hard core" of contents, which starts to be established in the nineteenth century. This "hard core" consists of the following groups of contents:

- "Localization and Orientation", particularly the content that refers to "Direction/ Orientation";
- "Scale", particularly the content that refers to "Cartographic scale";
- "Coordinates and imaginary lines", particularly the contents "Shape of the Earth/Movement of celestial bodies", "Hemispheres", "Imaginary lines/parallels and meridians", "Latitude and Longitude" and "Time Zones";

**Table 6.4** Cartographic contents (concepts, notions and subjects)

| | Geometric shapes | | Interpretation of maps |
|---|---|---|---|
| **Localisation and Shapes Orientation** | Shape of objects | **Cartographic representations (two-dimensional)** | Plan |
| | Symbols | | Map |
| | Legend | | World Map |
| | Points of view | | Sorts of projections |
| | Mind Map | | Topographic map |
| | Localisation | | Outline |
| | Laterality | | Sketch |
| | Direction / Orientation | | Thematic maps |
| **Scale** | Proportion | | Choropleth map |
| | Measurement | | Anamorphosis |
| | Cartographic scale | | Graphs and diagrams |
| | Geographical scale | **Cartographic representations (three-dimensional)** | Scale model |
| **Coordinates and imaginary lines** | Coordinates (elementary notions) | | Block diagram |
| | Shape of the Earth / Movement of celestial bodies | | The Globe |
| | Hemispheres | **Technology and Cartography** | Aerial photographs |
| | Imaginary lines / parallels and meridians | | Satellite images |
| | Latitude and Longitude | | GIS |
| | Time Zones | | Cartographic techniques |
| | History of Cartography | | |

- "Cartographic representations (two-dimensional representations)", particularly the content that refers to "Map";
- And "Cartographic representations (three-dimensional representations)", particularly the content that refers to "The Globe".

As it is possible to notice, the mentioned contents establish a continuity, a "school tradition" that dictates certain cartographic notions that "must" be worked in the first year of secondary education for almost two centuries, lasting for practically all the curricular reforms and the programmes of textbooks produced during the twentieth century and also permeating the twenty-first century.

Another important aspect that we perceived with the analysis refers to the work done with the notions of "Symbols" and "Legend". Even though some textbooks dealt with them during the thirties and forties, these contents reappear as prescriptions in all consulted textbooks that were published from 1972 to 2002. At this moment, there is a need to carry out some parallelisms between the domain of school knowledge and the domain of production of academic knowledge. We understand that working with these subjects has become more powerful due to the discussion in the academic sphere about the systematization of thematic cartography which is specially reinforced by the studies of Semiology of Graphics developed by Professor J. Bertin, from the end of the sixties to the beginning of the seventies (Martinelli 2000, 2007). This parallelism points to an educational transposition from this academic knowledge to a knowledge to be taught, prepared especially by authors of textbooks. On the other hand, the development of learning activities directed to the interpretation of maps, which is recommended by the last three curricular programmes analyzed, is not a procedure that is transposed to textbooks.

However, as a general rule, we can notice a process, from the seventies on, that makes Cartography subjects become more complex and the contents prescribed by the curricular programmes and textbooks, during the nineties, more comprehensive. We can see again the process of debates within the academic domain reflecting on the work of professors and authors of official curricula and teaching materials. This is because, right during this period, more accurate researches on how to teach Cartography were conducted in Brazil. At first they occurred by means of studies of few researchers, like the case of pioneering researches developed by Professor Lívia de Oliveira, from Universidade Estadual Paulista (UNESP), in Rio Claro, during the seventies (Almeida 2007).

However, from 1995 on, for the first time in Brazil, it was established a forum for debates concerning Cartography and its implications in teaching Geography, called "Colloquies of Cartography for Children", which are held approximately every 2 or 3 years. The colloquies became, therefore, a way to exchange information and to spread researches through their annals. The academic production in this field of research has also been spread and featured prominently in periodicals and newspapers, becoming the core for discussions in important events, like in the National Meeting of Geography Teaching (ENEG) and the Meeting of Geography Teaching Practice (ENPEG). Thus, it is possible to notice in the last decades, a process of transposition of this academic production to a knowledge to be taught through the curricular programmes and teaching materials.

## 6.5 Conclusions

According to what we observed a great part of the group of geographical concepts and, more specifically, of the geographical concepts prescribed in the national teaching materials, doesn't have its origins in a systematic academic knowledge.

We verified that its origins are supported by a sort of classical and erudite knowledge, based on the spirit of humanities, which formed the foundation for the secondary education in our country, before the establishment of the first universities.

Also, much of this classical geographical knowledge and, as a result, Cartography is part of it, is still present at schools by means of what we call "hard core of cartographic contents": a set of notions, concepts and subjects such as "Direction and Orientation", "Shape of the Earth and Movement of celestial bodies", "Imaginary lines: parallels and meridians", "Geographic coordinates/Latitude and Longitude", "Map" and "Globe", that have lasted in the Brazilian curriculum of Geography addressed to secondary education for the last two centuries, approximately. These explicit concepts, as well as the teaching method historically established by teachers-authors of textbooks, whom most of them also contributed to the production of the official programmes, at least until the forties, demonstrate a distinguished cultural production in which we verified that school Geography appears not only as a vulgarization or an adaptation of scientific Geographic knowledge, but as a characteristic and original knowledge from the educational institution for the educational institution. Thus, two different groups of knowledge were set up: the one intended for teaching and another for the academy.

## References

Almeida RD (2003) Cartografia na escolar. http://www.tvebrasil.com.br/salto/boletins2003/ce/index.htm. Accessed 3 May 2005

Almeida RD (2007) Apresentação. In: Almeida RD (ed) Cartografia escolar. Contexto, São Paulo

Almeida RD, Sanchez MC, Picarelli A (1997) Atividades cartográficas, vol 2, 3 e 4. Atual, São Paulo

Bittencourt CMF (1993) Livro didático e conhecimento histórico: uma história do saber escolar. Dissertation, Universidade de São Paulo

Brazilian Department of Elementary Education (1998) Parâmetros curriculares nacionais: geografia. MEC/SEF, Brasília

Callai HC (1999) A geografia no ensino médio. Terra Livre 14:60–99

Colesanti MTM (1984) O ensino de Geografia através do livro didático no período de 1890 a 1971. Dissertation, Institute of Geosciences and Exact Science, Universidade Estadual Paulista

Gasparello AM (2002) Historiografia didática e pesquisa no ensino de história. In: X encontro regional de história, ANPUH-RJ, "História e Biografias". Universidade do Estado do Rio de Janeiro

Gasparello AM (2006) Traduções, apostilas e livros didáticos: ofícios e saberes na construção das disciplinas escolares. In: Usos do passado, XII encontro regional de história ANPUH-RJ. Universidade do Estado do Rio de Janeiro

Issler B (1973) A geografia e os estudos sociais. Doctorate. Faculty of Philosophy, Science and Arts, São Paulo

Le Sann JG (2005) A caminho da Geografia: uma proposta metodológica. Dimensão, Belo Horizonte

Martinelli M (2000) As representações gráficas da Geografia: os mapas temáticos. Dissertation, Faculty of Philosophy, Arts and Human Science, Universidade de São Paulo

Martinelli M (2007) A sistematização da cartografia temática. In: Almeida RD (ed) Cartografia escolar. Contexto, São Paulo, pp 193–220
Simielli ME (1993) Coleção primeiros mapas. Ática, São Paulo
Simielli ME (1996) Cartografia e ensino: proposta de contraponto de uma obra didática. Dissertation, Faculty of Philosophy, Arts and Human Science, Universidade de São Paulo
Simielli ME (2008) Cartografia no ensino fundamental e médio. In: Carlos AF (ed) A geografia na sala de aula. Contexto, São Paulo, pp 92–108
Vechia A (2007) Os livros didáticos de história do Brasil na escola secundária brasileira: a produção dos saberes pedagógicos no século XIX. In: Proceedings of the simpósio internacional livro didático: educação e história, São Paulo
Vechia A, Lorenz KM (1998) Programa de ensino da escola secundária brasileira: 1850–1951. Authors' publisher, Curitiba

# Chapter 7
# The Transition from the Analytic to Synthesis Reasoning in the Maps of Geographic School Atlases for Children

**Marcello Martinelli**

**Abstract** At the end of nineteenth century, geographic school atlases gained credit among didactic materials. The "Atlas do Império do Brazil" of Almeida was the first Brazilian school Atlas. Currently there are many geographic school atlases on a variety of formats. Elaborating a school atlas is not a simple task. The first step for its coordination is the integrated interlacement of two fundamental orientations: the teaching of the map, based on the construction of the space notion by the student and its representation; and the teaching through the map, founded on the knowledge of the world from the nearby—the place—to the distant—the worldwide space. Regarding thematic maps, the representation of the reality can be carried out within analytic or synthesis reasoning. Analytic maps represent places, paths or areas characterized by attributes or variables. Synthesis maps identify groupings of places, paths or areas characterized by groupings of attributes or variables.

## 7.1 Opening

In the teaching and learning environment of Geography since its institution as a school subject, first in Germany and after in France, on the second half of the nineteenth century, geographic school atlases gained credit among didactic materials, increasingly adapting themselves to this function in the classroom.

They have appeared as selections and simplifications of the great general reference atlases, which evolved from simple collections to a systematic organization with specific and intellectual purpose.

M. Martinelli (✉)
Departamento de Geografia—Pós-graduação, Programa Geografia Humana, Faculdade de Filosofia, Letras e Ciências Humanas—Universidade de São Paulo, São Paulo, SP, Brazil
e-mail: m_martinelli@superig.com.br

L. Zentai and J.R. Nunez (eds.), *Maps for the Future*, Lecture Notes in Geoinformation and Cartography 5,
DOI 10.1007/978-3-642-19522-8_7, 

The "Atlas général Vidal-Lablache: histoire et géographie" of 1824 was a classic that inspired several derivations in France as well as in other countries of the continent.

In 1868, the first Brazilian school atlas was published, called "Atlas do Império do Brazil" of Cândido Mendes de Almeida. It was adopted by the Colégio Imperial Pedro II, in Rio de Janeiro.

As a result of the whole evolution and epistemological transformation of atlases cartography, a great number of assorted geographic school atlases are currently available in print and electronic formats.

## 7.2 Elaborating a Geographical School Atlas

This elaboration is not a simple task. It will have to start with the lucubration about the construction of the space notion and its representation by the student. As chief sources, among others, there are the psychogenetic studies of Jean Piaget and his team and of other researches, like the ones from Vygotsky and Wallon (Piaget and Inhelder 1972; Wallon 1995; Vygotsky 1998).

In Brazil, we count on the contributions of Doctor Professor Lívia de Oliveira, who established the master lines for a correct orientation of these works, having instituted a proper school with high qualified disciples (Oliveira 1978, 2006; Almeida and Passini 1989; Almeida 2001).

A school atlas will not be only a collection of maps, ready and finished, but a systematic organization of representations developed with a specific intellectual purpose. With this intent, the articulation of two fundamental methodological bases has to be taken into account: the methodological basis of the map and the methodological basis of the acquisition of knowledge in geography trough the map.

Therefore, maps would not be seen as mere illustrative figures of didactic texts, but as representations that reveal questions that will be tackled and discussed within the geographic speeches, opening space to critical and conscious reflections.

The atlas enterprise considers the integrated interlacement of two basic orientations as a first step for its coordination:

- The teaching of the map, propagated on the theorist-methodological postures about the construction of the space notion and its respective representation by the scholar, involving initial cartography practices,
- The teaching through the map, accomplished on geography, promoting the knowledge of the world from the spatial inclusion and continuity, starting by the nearby, experienced and well known—the place—to the distant unknown—the worldwide space.

The atlas content is organized in consonance with the context of the geography methodological bases.

The spatial outline definition, ranging from the local place to the worldwide space, would come linked to the question of the thematic and scale structure.

Afterwards, the atlas Thematic Cartography is based upon the map elaboration as a construction within the parameters that consider the graphic representation as a language, integrating a monosemic semiotic system (having only one meaning) (Bertin 1973, 1977).

Within this context, the atlas thematic maps can be constructed by selecting the most appropriate method to the characteristics and forms of manifestation (on dots, on lines, on areas) of the phenomena of the reality took into account on each theme, following a qualitative, ordered or quantitative approach.

The representations can also undertake either a static or dynamic appreciation of the reality. Yet, the phenomena that compose the reality to be represented on a map could be glimmered within analytic or synthesis reasoning. Thus, on the one hand, there is an attention to the constitutive elements, the places, paths or areas characterized by attributes or variables but on the other hand, there are integrated spatial units, which mean groupings of places, paths or areas being characterized by groupings of attributes or variables.

## 7.3 The Analytic Reasoning

The analytic reasoning in maps is directed to the scrutiny of the geographic space, mobilizing procedures of classification, combination and search for explanations about facts or phenomena seen indistinctly on the reality. The mental operations undertaken, regarding analytic maps, will allow the student to formulate conjectures about what would elucidate the phenomena geography. However, before a more rigorous critic, authors affirm that they would not be able to suggest the causalities or give the explanations by themselves, but to indicate new researches (Rimbert 1968; Claval and Wieber 1969).

Analytic maps are the most widespread on school atlases. They can display the representations on a qualitative, ordered or quantitative approach, considering manifestations on points, lines, and areas, according to the static or dynamical point of view.

Qualitative maps express the existence, the location and the extension of the events on a certain situation on the time, that are distinguished by their nature, species, allowing them to be classified by criteria established by the sciences that study them.

Ordinate maps show, on a certain date, categories that are enlisted on a sole sequence, defining hierarchies, or focalize, on a single map, aspects that were being consolidated over the time.

Quantitative maps evidence the relationship of proportionality between quantities that characterize places, paths or areas for a certain moment. Look at the following example: the map shows in analytic form the ternary structures of Brazil land use of the establishment total areas in 2006 (Fig. 7.1).

From a dynamical point of view, maps can show qualitative, ordered and quantitative variations on the time or qualitative, ordered and quantitative directed movements on the space.

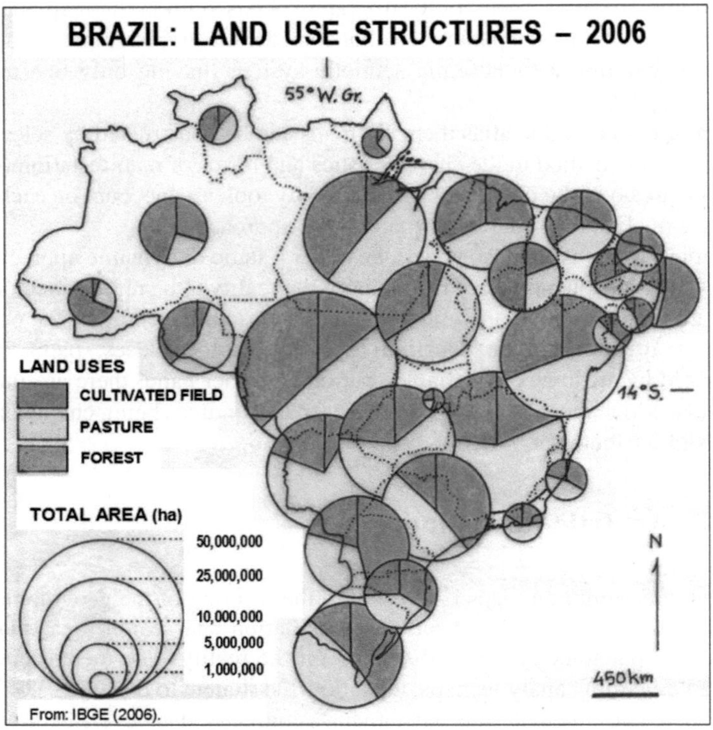


**Fig. 7.1** The distribution of ternary structures in the country

## 7.4 The Synthesis Reasoning

Synthesis maps would have, as their first function, pointing out the correlations, evidencing connections between distinct phenomena (Claval and Wieber 1969).

Synthesis is a necessity, but it has to be treated in a way that makes new configurations to emerge, completely different from those resulted of a simple sum of the elementary configurations. Just in this manner, an overview of the reality would be achieved.

These maps become a privileged instrument of the geographer, who is interested in regional studies.

Despite this entire methodological base, established with the evolution of the cartographic science, it is noticeable that still exist a lot of confusion about what synthesis cartography is. And this is transferred to school atlases.

Many people still conceive it, by means of maps stated—synthesis maps—not like logical systems, but like superposition or juxtaposition of analyses. Then, very confused maps appear, on which a lot of symbols are accumulated, denying the very idea of synthesis.

On synthesis, we cannot have the elements on superposition or juxtaposition anymore, but their fusion in "types". This means that, regarding maps, it is possible to identify groupings of places, paths or areas—as elementary spatial units of observation—characterized by groupings of attributes or variables (Rimbert 1968; Bertin 1973, 1977; Martinelli 1998, 2003, 2005, 2011b, Martinelli and Ferreira 1995).

To explain what synthesis reasoning is, it was borrowed an experimental piece of work done by Gimeno (1980). Its aim was to discover which groupings could be formed in a set of 42 elementary data: seven objects related to six attributes. The following diagram shows the transition of the analytic moment; where, in a matrix, each object relates to one or more attributes; to the synthesis moment, achieved with reiterated permutations between columns and rows of the matrix, revealing three groups of objects characterized by three groups of attributes. This is the revealed information (Fig. 7.2).

Synthesis cartography can be done by an assorted gamma of methods that were being developed together with an accurate search, involving qualitative, ordinate or quantitative data, on a static or dynamical appreciation, by employing entities like points, lines and areas.

Thus, according to the objectives and established fields of study, basically three main groups can be pondered: traditional superposition and manual combination of many analytic thematic maps, graphic methods and statistic-mathematic methods.

The triangular graph method will be considered here due to its intrinsic simplicity. It is employed on a particular case in synthesis cartography, the one which searches for the representation of the "types" of specific ternary structures on space, that is, for variables formed by three collinear components. In this way, such

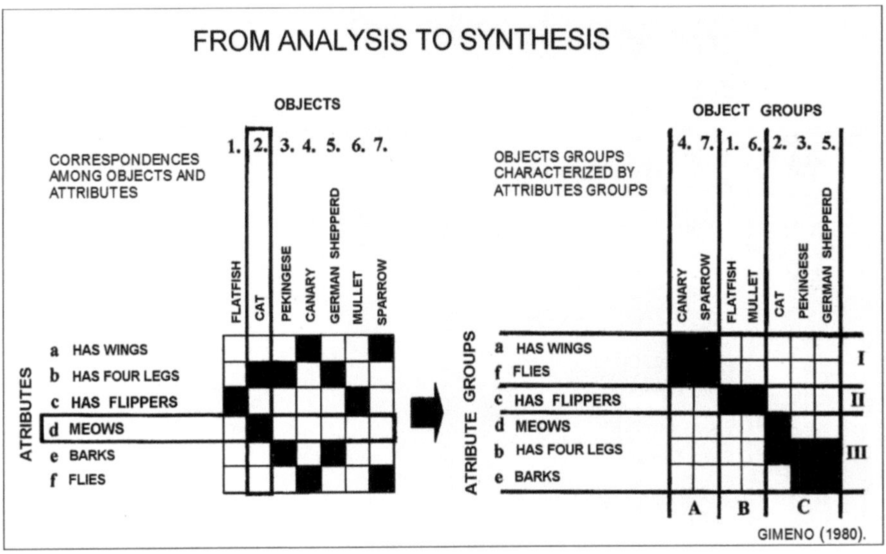


Fig. 7.2 The transition from analytic to the synthesis reasoning

graphic would participate as an algorithm for the treatment of the data and for the legend organization.

The different combinations of the three components I, II and III of the studied variable are synthesized through dots inside the triangle. Since the variable refers to areas, each dot of the graphic represents the structure of each one (Beguin and Pumain 1994).

Based on the visual analysis of the resultant point pattern, the areas are grouped according to categories defined by the position they assume in the triangle. Sometimes the groupings are not that easy to be discerned. A more accurate control is required. The categories, defined as such, will be later transferred to the map, which will represent the synthesis of the ternary structures grouped into significant classes. The triangular graph will be its legend, giving total transparency to the reasoning undertaken on the map construction.

The following representation shows the synthesis of the Brazil information displayed in the Fig. 7.1 as "types" of land use structures (Fig. 7.3).

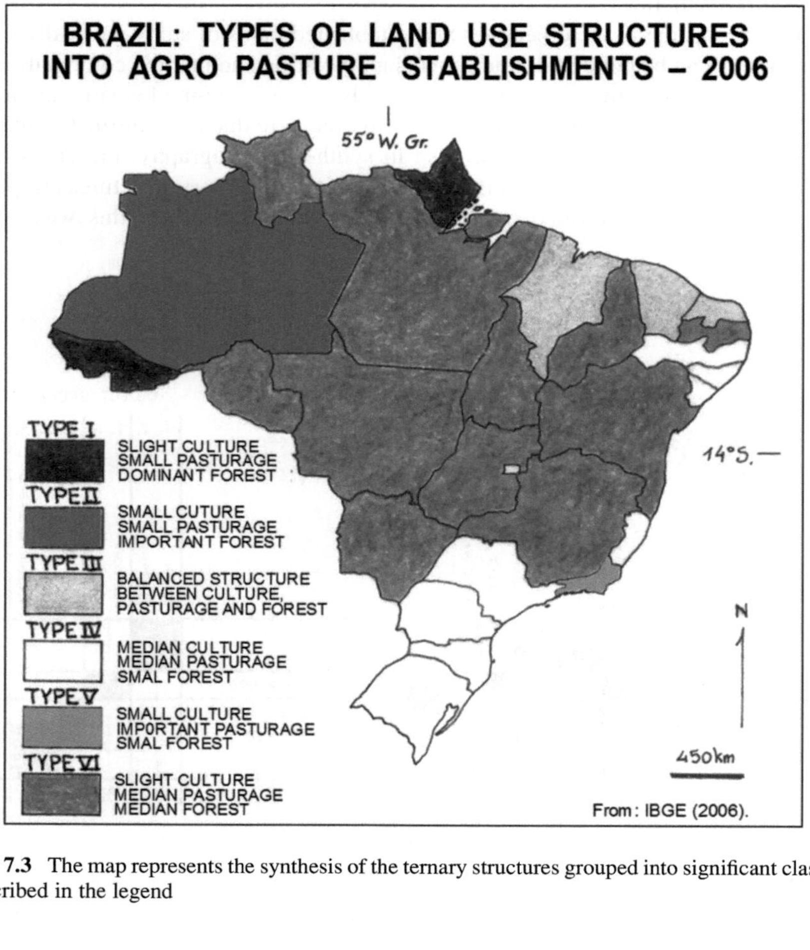


**Fig. 7.3** The map represents the synthesis of the ternary structures grouped into significant classes described in the legend

Generally, synthesis cartography worked by graphic methods, is explored along with static situations. However, it is also possible to elaborate it in dynamical approaches like in the case of the establishment of demographic evolution "types" for a certain period of time with various census data.

To get to this synthesis it is mobilized a graphic treatment of the data, which consists in elaborating an evolution graph in semi-logarithmic graph-paper for each observation unit. After ready, these diagrams will be visually classified, approximating those that are more similar, trying to form groups with similar evolutionary characteristics. Each group will consist of a "type" that will be qualified on the legend by a symbol and its respective epithet, expressed in a concise manner.

Each rubric of the legend, thus specified, will receive either an indicative colour or texture to be plotted on the map that will express the synthesis.

A remark on this topic is extremely important for school atlases. It is insufficient to use analytical and synthesis maps only to show the existence of these two kinds of reasoning. A consistent presentation of the synthesis map is only valid when the analytical maps, employed to accomplish the synthesis representation, are available to the students.

## 7.5 Final Remarks

These methodological directions are considered imperative to sustain all and any undertaking directed to the geographic school atlases idealization, when dealing with analytic and synthesis maps. They will confirm to them, in consistently form, their pedagogic role in geography, preparing the citizen for the practice of social transformation.

## References

Almeida RD (2001) Do desenho ao mapa: iniciação cartográfica na escola. Contexto, São Paulo
Almeida RD, Passini EY (1989) O espaço geográfico: ensino e representação. Contexto, São Paulo
Almeida RD et al (1997) Atividades cartográficas. vol 4. São Paulo, Atual
Antunes AR et al (1993) Estudos sociais: teoria e prática. Access, Rio de Janeiro, Access
Beguin M, Pumain D (1994) La représentation des données géographiques: statistique et cartographie. Armand Colin, Paris
Bertin J (1973) Sémiologie graphique: les diagrammes, les réseaux, les cartes. Mouton, Gauthier-Villars, Paris
Bertin J (1977) La graphique et le traitement, graphique de l'information. Flammarion, Paris
Claval P, Wieber JC (1969) La cartographie thématique comme méthode de recherche. Les Belles Lettres, Paris
Gimeno R (1980) Apprendre à l'école par la graphique. Retz, Paris
Martinelli M (1998) Mapas e gráficos: construa-os você mesmo. Moderna, São Paulo
Martinelli M (2003) Atlas geográfico: natureza e espaço da sociedade. Editora do Brasil, São Paulo

Martinelli M (2005) A student geographic atlas for the natural and social spaces learning. In: Proceedings of the ICA ICC 2005, A Coruña
Martinelli M (2011a) Atlas geográficos para escolares: uma revisão metodológica. In: Almeida RD (ed) Novos rumos da cartografia escolar: currículo, linguagem e tecnologia. Contexto, São Paulo
Martinelli M (2011b) Os mapas da geografia e cartografia temática. Contexto, São Paulo
Martinelli M, Ferreira GML (1995) L'atlas géographique illustré: un premier atlas pour les enfants. In: Proceedings of the ICA ICC 1995, Barcelona
Oliveira L (1978) Estudo metodológico e cognitivo do mapa. USP-IG, São Paulo
Oliveira L (2006) Os mapas na geografia. Geografia 31(2):219–239
Piaget J, Inhelder B (1972) La représentation de l'espace chez l'enfant. PUF, Paris
Rimbert S (1968) Leçons de cartographie thématique. Sedes, Paris
Vygotsky LS (1998) A formação social da mente. São Paulo, Martins Fontes
Wallon H (1995) A evolução psicológica da criança. Edições 70, Lisboa

# Chapter 8
# How School Trips Modify the Pupils' Representation of Space

**Katarzyna Bogacz**

**Abstract** A research project studies the modalities of spatial learning. The objective was to understand if the experience of school trips modifies the children's spatial representations. The research was carried out with 192 pupils in Lyon (France) and Cracow (Poland). In order to examine the children's spatial representations, mental maps were used. In this method, the geographical approach places the lived-in space in the centre of preoccupations. The research postulated that the connection to reality is inseparable from the filter of the representations. Through experiences, the individual constructs an interior model of his or her environment. This paper reports specifically on the results of a sample study in Poland.

## 8.1 The Theoretical Framework

My work is related to two disciplines, geography and psychology. It is supported by the postulate of spatial representations within the framework of the paradigm of spatial production. The theoretical and epistemological reflections of geographers, which consider the values attributed by the individual to the space, lead to mobilizing the theoretical and methodological elements of psychology. The representations, in the sense of reconstitution and interpretation, are very important in this teaming of disciplines. Looking at space as representation seems to be triple facetted: according to Berdoulay (1988, cited by Di Méo 1991), the geographical space can be understood as a complex construction where the subject, the spatial reality of the earth, and the subject's representations intervene.

K. Bogacz (✉)
Lumière University Lyon 2, Géography Research Institut (IRG-UMR 5600), Bron cedex, France
e-mail: k.bogacz@univ-lyon2.fr; k.bogacz@hotmail.fr

L. Zentai and J.R. Nunez (eds.), *Maps for the Future*, Lecture Notes in Geoinformation and Cartography 5,
DOI 10.1007/978-3-642-19522-8_8, 

## 8.2 Presentation of the Research Protocol

### *8.2.1 Research Objectives*

One of the central questions of this research project is to answer the following question: how does the school trip modify the children's spatial representations? This key question raises another question, which asks about the tools, namely, how to understand the "structure" of spatial representation? Another question asks about the materials: how to identify the structure of children's spatial representation of their urban environment in Lyon and Cracow? At the moment, it is possible to answer partially the initial question and to evaluate the effects of spatial mobility on representations, particularly on their structure.

### *8.2.2 Research Hypotheses*

The following hypotheses were put down:

- H1: The spatial experience as a result of the school trip of a few days modifies the children's urban spatial representations in terms of their level of structure.
- H2: Children's urban spatial representation is more structured for those who have already benefited from the personal experience of mobility.
- H3: The field trip modifies the children's level of spatial structure differently according to their personal experience of mobility.
- H4: The field trip modifies the children's level of spatial structure differently according to their school experience of mobility.

### *8.2.3 Field Research*

My research is based on a specific field: the school classes that went on 5-day school trips (called "discovery classes" in France and "green schools" in Poland) during the spring of 2008. I carried out the work in Lyon and Cracow at the same time.

The object of my research is primary schools. I chose classes that would allow me to constitute a data corpus that presents a high degree of homogeneity (same age, belonging to a common territory). In Lyon, I carried out the survey in five school classes (121 pupils) of the same level from three primary schools. In Cracow, I collected my data with the pupils of five classes at three different levels from three primary schools.

The sample of the children in my research was formed on the basis of the following measures: the age (from 9 to 12), and the length and the kind of the school trip. The length of the 5 days was chosen principally for the possibility of comparing the two research fields. The benefits expected from the school trip vary in the function of their length. Thus, according to a regulatory text effective in France, the discovery classes with the length equal to or longer than 5 days allow the pupils to escape significantly from the usual context and space of the class. Thus, they constitute the actual change of scenery and the privileged moment of collective life apprenticeship. As there is a great number of "discovery classes", I thought it very important to concentrate my research only on one kind of school trip. As Giolitto (1978) argues, the educative quality of "green classes" seems to be superior to that of other "discovery classes", because the true contact with nature is more formative than at a simple trip of skiing or horse-back riding. This explains why I chose "green classes".

I constructed my sample of pupils from five classes at Cracow and Lyon. Their ages ranged from 9 to 12. All these children went on a 5-day school trip in the period from 19 May to 20 June 2008. In order to be sure that the changes observed in the children are the results of their stay in the "discovery classes", I conducted my survey in the shortest possible time before and after the school trip.

### *8.2.4 Research Methodology*

I collected two kinds of data: graphic (children drawings) and discursive (questionnaire for the children)—before and after the school trip. I started with the questionnaire in order to identify the age, the origin, the place of residence and the sex of the children, but also their experience of mobility (number of departures on holiday per year, destinations, previous experience in discovery classes). In the next step, I wanted to obtain their mental map, which is defined by Bailly (1985) as the product that is the individual representation of spatial environment. According to Bailly, the mental map permits the subject to fix the images of his or her environment and find the limits of spatial knowledge. As a result, this approach allows us to comprehend individual and collective representations of space while allowing each subject a great amount of freedom in their manner of expressing themselves. Subsequently, after the school trip, I asked the children to draw one more mental map and to draw their city on a white paper, without documents or oral supplementary indication. According to Gumuchian (1988), drawing constitutes the material able to translate the space representation of a child. It is a child friendly means of expression: drawing can be particularly significant and permit a formulation more spontaneous and direct than writing. Because of the different conditions of collecting our research materials (ten classes in two different education systems), I made the most precise translation to be able to compare them. I personally conducted each survey and I asked the teacher who was present during the research not to give any information to the pupils. As I carried out my research in two cities using two different languages, I took care of giving the simplest and the shortest possible instructions.

### 8.2.5 *Mental Map's Exploitation*

Mental maps analysis opens several routes of reflection. I concentrated on the reading of the interpretation, which could account for (and compare) different ways of space representation. The question is to elaborate the model of analysis and interpretation of mental maps, and this one is, obviously, foreign to classic topographical representation. Thus, I used the matrix of Kevin Lynch. According to Lynch, the images could be distinguished by the quality of their structure.

K. Lynch (1998) proposed four degrees of the mental maps' structure:

1. No structure
   The various elements are unlinked. There is neither structure nor interrelation between the elements (Fig. 8.1).

**Fig. 8.1** No structure

2. Structure of position
   The parts are separated but linked up to a certain extent in terms of their general direction and possibly the distance (Fig. 8.2).
3. Flexible structure
   The parts are linked up but in a flexible way (Fig. 8.3).
4. Rigid structure
   The parts are linked in all directions (Fig. 8.4).

## 8.3 Research Results

### *8.3.1 Presentation of a Sample from Poland*

Tables 8.1 and 8.2 present a sample of the Polish survey (three schools and three different levels) from the aspect of the research hypothesis: the "previous stay in a discovery class" and "frequency of departures on holidays".

#### 8.3.1.1 Quantitative Treatment of Data

Next, I present the part of the research that concerns the pupils from the same school of Cracow. The pupils are of the same age and from the same class. Their space belonging is nearly the same.

Table 8.3 shows the structure of mental maps made by the pupils from Cracow (9–10 years old). It can be easily observed that all of them changed. The main difference is the "non structure"—after a 5-day school trip, there are less spatial representations with unlinked elements. At the same time, after the "green school", there are more "flexible structure" and "rigid structure".

As Table 8.4 shows, 9–10 years old pupils from Cracow who go on holidays at least three times a year had principally spatial representations with "non structure" (34.6%) or "flexible structure" (34.6%) before the school trip. After the "green school", there is more "structure of positions" (30.8%). "Flexible structure" is less than before the school trip, and at the same time, "rigid structure" goes up from 7.7% to 19.2%.

Before the school trip, the structure of nearly 90% of mental maps was "non structure" or "structure of positions" for 9–10 year old pupils in the sample from Cracow who go on holidays less than three times a year. After the school trip the difference is significant—55.6% of mental maps with "flexible structure". However, as Table 8.5 shows, this change does not concern "non structure", but only "structure of positions".

The number of spatial representations of pupils with previous experiences on "discovery classes" is exactly the same for "structure of positions" and "flexible structure" before and after the school trip (Table 8.6). At the same time, there is less "non structure" and more "rigid structure" after the "green school".

**Fig. 8.2** Structure of position

**Fig. 8.3** Flexible structure

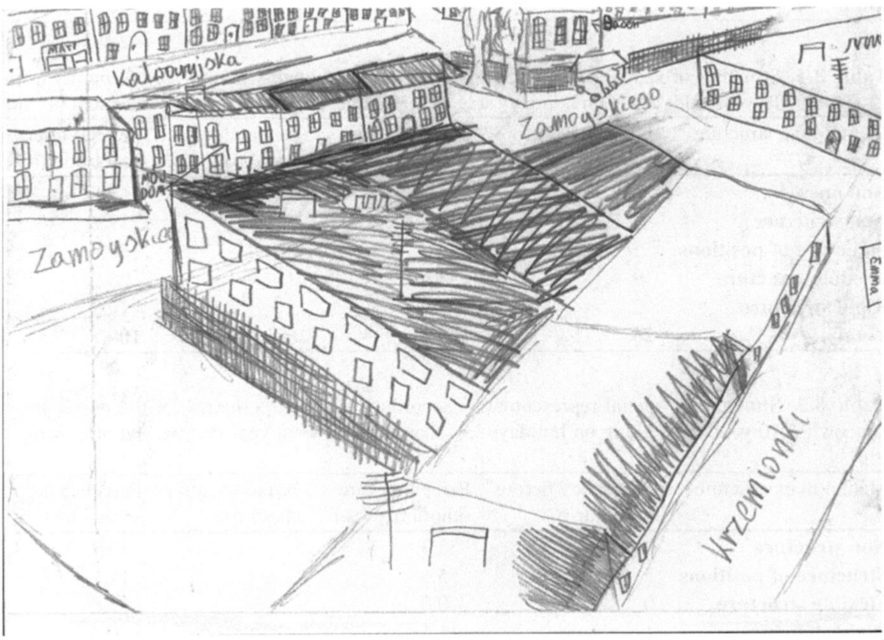


**Fig. 8.4** Rigid structure

**Table 8.1** Pupils from Cracow and their "previous discovery classes"

| Class/age/school | Non answer | Yes | No | Total |
|---|---|---|---|---|
| **III (9–10) School 1** | 1 | 17 | 17 | 35 |
| **IV (10–11) School 2** | 0 | 16 | 0 | 16 |
| **V (11–12) School 3** | 0 | 14 | 6 | 20 |
| **Total** | **1** | **47** | **23** | **71** |

**Table 8.2** Pupils from Cracow and their "frequency of departures on holidays"

| Class/age/school | Less than 3 times/year | At least 3 times/year | Total |
|---|---|---|---|
| **III (9–10) School 1** | 9 | 26 | 35 |
| **IV (10–11) School 2** | 6 | 10 | 16 |
| **V (11–12) School 3** | 5 | 15 | 20 |
| **Total** | **20** | **51** | **71** |

**Table 8.3** Structure of spatial representations according to Lynch's matrix for the pupils from Cracow (9–10 years old) before and after school trip

| Mental maps' structure | Frequency before school trip | Percent before school trip (%) | Frequency after school trip | Percent after school trip (%) |
|---|---|---|---|---|
| **Non answer** | 1 | 2.9 | 1 | 2.9 |
| **Non structure** | 12 | 34.3 | 8 | 22.9 |
| **Structure of positions** | 10 | 28.6 | 9 | 25.7 |
| **Flexible structure** | 9 | 25.7 | 12 | 34.3 |
| **Rigid structure** | 3 | 8.6 | 5 | 14.3 |
| **Total** | **35** | **100** | **35** | **100** |

**Table 8.4** Structure of spatial representations according to Lynch's matrix for the pupils from Cracow (9–10 years old) going on holidays at least three times a year (before and after school trip)

| Mental map structure | Frequency before school trip | Percent before school trip (%) | Frequency after school trip | Percent after school trip (%) |
|---|---|---|---|---|
| **Non answer** | 1 | 3.8 | 1 | 3.8 |
| **Non structure** | 9 | 34.6 | 5 | 19.2 |
| **Structure of positions** | 5 | 19.2 | 8 | 30.8 |
| **Flexible structure** | 9 | 34.6 | 7 | 26.9 |
| **Rigid structure** | 2 | 7.7 | 5 | 19.2 |
| **Total** | **26** | **100** | **26** | **100** |

**Table 8.5** Structure of spatial representations according to Lynch's matrix for the pupils from Cracow (9–10 years old) going on holidays less than three times a year (before and after school trip)

| Mental map structure | Frequency before school trip | Percent before school trip (%) | Frequency after school trip | Percent after school trip (%) |
|---|---|---|---|---|
| **Non structure** | 3 | 33.3 | 3 | 33.3 |
| **Structure of positions** | 5 | 55.6 | 1 | 11.1 |
| **Flexible structure** | 0 | 0.0 | 5 | 55.6 |
| **Rigid structure** | 1 | 11.1 | 0 | 0.0 |
| **Total** | **9** | **100** | **9** | **100** |

**Table 8.6** Structure of spatial representations according to Lynch's matrix for the pupils from Cracow (9–10 years old) with previous experience of "discovery classes" (before and after school trip)

| Mental map structure | Frequency before school trip | Percent before school trip (%) | Frequency after school trip | Percent after school trip (%) |
|---|---|---|---|---|
| **Non structure** | 4 | 23.5 | 1 | 5.9 |
| **Structure of positions** | 6 | 35.3 | 6 | 35.3 |
| **Flexible structure** | 6 | 35.3 | 6 | 35.3 |
| **Rigid structure** | 1 | 5.9 | 4 | 23.5 |
| **Total** | 17 | 100 | 17 | 100 |

**Table 8.7** Structure of spatial representations according to Lynch's matrix for the pupils from Cracow (9–10 years old) without previous experience of "discovery classes" (before and after school trip)

| Mental map structure | Frequency before school trip | Percent before school trip (%) | Frequency after school trip | Percent after school trip (%) |
|---|---|---|---|---|
| **Non answer** | 1 | 5.9 | 1 | 5.9 |
| **Non structure** | 8 | 47.1 | 7 | 41.2 |
| **Structure of positions** | 4 | 23.5 | 2 | 11.8 |
| **Flexible structure** | 2 | 11.8 | 6 | 35.3 |
| **Rigid structure** | 2 | 11.8 | 1 | 5.9 |
| **Total** | **17** | **100** | **17** | **100** |

Many pupils without previous experience of "discovery classes", have "non structure" spatial representations (47.1% before and 41.2% after school trip). The principal change concerns the "flexible structure" after the "green school". "Structure of positions" and "rigid structure" change a little (Table 8.7).

## 8.4 Conclusions

The main conclusions show:

- Indication of personal experience: Urban spatial representations are more structured for those pupils who go on holidays at least three times a year.
- Indication of "discovery classes" experience: School trips modify the level of the pupils' spatial structure differently according to the pupils' personal or school experiences of mobility.

This study permits us to evaluate the effects of "discovery classes" as a spatial learning experience, together with the family background. The analysis brings more concrete elements to the global problem of "spatial capital" (Lévy and Lussault 2003) and its individual transmission and development.

## References


Bailly A (1985) La région: de la territorialité vécue aux mythes collectifs. In: Lajugie J (ed) Région et aménagement territoire. Brière, Bordeaux

Berdoulay V (1988) Des mots et des lieux. Editions du Centre national de la Recherche Scientifique, Paris

Di Méo G (1991) L'Homme, la Société, l'Espace. Anthropos, Paris

Giolitto P (1978) Classes de nature. Casterman, Paris

Gumuchian H (1988) De l'espace au territoire, représentations spatiales et aménagement. Université Joseph Fournier, Grenoble

Lévy J, Lussault M (eds) (2003) Dictionnaire de la géographie et de l'espace des sociétés. Belin, Paris

Lynch K (1998) L'image de la Cité. Dunod, Paris

# Chapter 9
# Interpretation of Surface Features of Mars as a Function of Its Verbal—Toponymic—and Visual Representation

**Henrik I. Hargitai**

**Abstract** Does the language of place names affect the interpretation of a planetary map? Is there any consistent change in the reading of a planetary map if the language of descriptor terms is changed from the official Latin to the mother language of the map reader? A survey made at a middle school in Hungary addressed these questions. The method was to collect preconceptions the landscape of Mars, then to show maps in Latin and Hungarian, and again ask the same questions. The results show that there are considerable differences in the two groups.

## 9.1 Introduction

Names of the surface features on extraterrestrial planetary bodies (planets, moons, minor planets, dwarf planets, cometary nuclei) are approved by the International Astronomical Union (IAU) Working Group for Planetary System Nomenclature (WGPSN) and listed in the Gazetteer of Planetary Nomenclature. Usually, names are binominal, having a "designation" (e.g. "Olympus"), and a "descriptor term" (e.g. "Mons") (always in Latin), except for craters and some other features, which have no descriptor term (USGS 2011; Hargitai 2006).

One of the projects of the Commission on Planetary Cartography of the International Cartographic Association is to develop planetary maps for the general public and develop standardized national variants of the official Gazetteer of Planetary Nomenclature (Hargitai et al. 2010; Hargitai et al. 2008; Hargitai and Kereszturi 2002, 2010; Shingareva et al. 2005).

The aim of *this* work is to investigate how the *language of the nomenclature* changes the perception of an unknown surface.

H.I. Hargitai (✉)
Department of Physical Geography, Eötvös Loránd University, Budapest, Hungary
e-mail: hhargitai@gmail.com

L. Zentai and J.R. Nunez (eds.), *Maps for the Future*,
Lecture Notes in Geoinformation and Cartography 5,
DOI 10.1007/978-3-642-19522-8_9, © Springer-Verlag Berlin Heidelberg 2012

The hypothesis was that (1) first consulting a planetary map with a nomenclature will provide a basis for a more detailed cognitive map than all previous knowledge on that planet and (2) if the opaque descriptor terms approved by the IAU WGPSN are translated to a language transparent to its reader, the map reader who gets a translated map will have a more realistic (cognitive) view of the surface than those map readers who did not understand the meaning of descriptor terms.

## 9.2 Theoretical Background

### *9.2.1 Phenomenology of Maps*

Cartography is today considered to be science, technology and art (for a detailed discussion see Krygier 1995). Its aesthetic aspect is (almost) never an art for art's sake, but a form of applied art or applied aesthetics. Although its main goal is to visualize spatial phenomena, it does this via generalization and interpretation; through the glasses of its makers and readers. Planetary geologic maps are generally based on photo*interpretation*. Photomaps (mosaics) have no human pre-interpretation process, they show the surface as they are—or as seen by a spacecraft—at a given (chosen or arbitrary) date, wavelength, resolution and illumination. In this case, the complex task of interpretation is passed to the map reader who is not necessarily as educated as needed for such performance. All kinds of maps show some degree of interpretation from both sides.

Since planetary maps are more or less standardized products and the common planetary map reader is a professional planetary scientist, therefore misinterpretation is unlikely. Map readers use these standardized maps for a very specific purpose: for navigation (orienteering, nautical, military, en-route, etc. charts) or for a scientific purpose (geologic, hydrologic maps, etc.). The case is different if a map—generally most small scale maps, and in our case: a planetary map—is produced for a general audience—non experts, students, or even children. The aim is not to find something fast and efficiently, but to browse on the map, to discover the represented surface—a landscape, which is not going to be or can't be visited, only by the map. These maps are popular science products: interpretations of scientific results. The goal is not to find something fast, but to have a good time while using the map. Here very little previous knowledge can be expected on this field therefore their map interpretation will be highly dependent on the (visual and textual) presentation of the data.

Polish philosopher Roman Ingarden (1974) proposes that the literary text contains gaps, "spots of indeterminacy", which the act of reading fills in, or "concretizes". This way the reader generates meanings not intended by the author. Although maps are not literary works, its "reception" works the same way: since no long explanatory texts and images on landforms and map units can be attached to general maps, the map readers will have to interpret symbols, textures, descriptor terms themselves, this way to generate meanings based on his or her terrestrial geographic experience modified by the preconceptions originating from fiction and non-fiction literature on the planet in question (this includes esoteric literature in many cases). I emphasize that this "concretizing" effect may be negligible for planetary scientists but may well suppress the intended message of the planetary cartographer in the case of the non-expert map readers. Philosophical phenomenology can't be applied to hard sciences, but any cartographic product is only a generalized (human) representation of the physical world, therefore *subjective experiences* in its perception are inevitable. Planetary maps—along with other maps—are human phenomena, based on (but not equal to) the physical reality, especially when considering their perception process.

Planetary nomenclature may be even more sensitive to cultural aspects since it mixes physical reality with social reality.

Nomenclature is part of the human language. One of Wittgenstein's main ideas is that "The propositions show the logical form of reality" (Wittgenstein 1922:4.121) In practice, place-names are only parts of the propositions ("thoughts with a sense") but in fact, a planetary place-name is also a full proposition which proposes that a feature (linked to reality—a specified place and area on a specified planet—by the Gazetteer) belongs to a particular geomorphic (and often, geologic) category. Applying this idea to the structure of descriptor terms (but not the specifics!) makes sense: descriptor terms show the logical form of reality.

Wittgenstein's famous "The limits of my language mean the limits of my world" (Wittgenstein 1922:5.6) statement can also be applied to planetary nomenclature. Its function is not only to be a "tool" as stated by the IAU Rules (USGS 2011) but also to expand human Oikumene by connecting (previously) alien worlds' surface features to the human world.

John Searle (1997) differentiates *brute facts* that can exist independently of humans and our institutions from institutional facts that require human institutions for their existence. He goes further, saying that "in order to state a brute fact we require the institution of language, but the fact stated needs to be distinguished from the statement of it". However, in planetary nomenclature we baptize physical objects and by doing this we connect them to our culture and also we delimit them (by defining their sizes) and classify them (according to our current geological understanding). Physical objects—or rather, their names and cartographic representations—become parts of our social reality when shown on maps labeled and defined. Brute facts are now inseparably connected to institutional facts by IAU's nomenclature and the chosen visual representation methods. In turn, they are mentally linked to science, high and pop culture, sci-fi universes, etc. Names (and certainly some categories) of planetary surface features exist only in the minds and mental maps of humans.

The perception of Geographical names is different when they are read from when they are heard. Reading a planetary geographical name is looking at its standardized symbol—a string of Roman characters, a label (Kadmon 2000). It is a typical example of artificial names that were born in written form. But still, they also become spoken words, parts of the oral tradition; and for many students they are heard first (at a lecture). In either case, letters appear as sounds when reading silently. "Oral folk have no sense of a name as a tag, for they have no idea of a name as something that can be seen. Written or printed representations of words can be labels, real, spoken words cannot be" (Ong 1982 [2002:33]). Thus, after pronouncing the name, its sound reminds us to the object itself, it brings the picture of the object to our minds—it is not just a representation, a label of it any more, as it is in its written form. Knowing a proper pronunciation is therefore just as important as the ability to read the names.

### 9.2.2 Ways Towards Standardized Planetary Place-Names

Latin names of the planetary nomenclature follow mediaeval geographers' and seventeenth to nineteenth century astronomers' now extinct naming conventions which were revitalized by twentieth century astronomers. For scientific purposes, the use of this official, standardized nomenclature is unquestionable.

The United Nations Group of Experts on Geographical Names (UNGEGN) defines international geographical names standardization as "an activity aimed at reaching maximum practical uniformity in the rendering ... of all geographical names on Earth." (UN 2006) The same principle can be applied for planetary surface features names as well, for international communication: it is also the only reasonable choice—it is universally searchable on the internet, which is important for finding data or picture galleries published in any languages.

However, for studying the surfaces by children, students, or the general public—especially in classroom situations—the translated variant of the generic parts (descriptor terms) and transcribed/transliterated variant of the specific parts may be more suitable, the latter especially in countries where they use a writing system other than the Roman alphabet. This conception is also confirmed by the recently growing number of the use of exonyms and phonetically transcribed place-names in school atlases (Jordan 2011) which trend has positive reception by the educators; a *domestic* trend in contrast to the *international* standardization efforts by UNGEGN (and IAU).

The WGPSN of IAU does not provide guidelines and does not encourage the creation of localized versions of the Gazetteer. One of IAU's naming rules state that "in general individual names chosen for each body should be expressed in the language of origin. Transliteration and pronunciation for various alphabets should be given but there will be no translation from one language to another" (IAU 1970). Despite this rule, there is no any transliterated forms given by the Gazetteer and it is not clear if this rule refers to planetary body names, surface feature names, specifics or descriptor terms. In the 2011 version, reference to "pronunciation" is left out of the rule (USGS 2011), and now IAU's standpoint is that "there are many variations on the pronunciation of names, and the IAU does not endorse any particular one" (personal communication with an undisclosed WGPSN member, 2010.10.06). Knowing a preferred pronunciation (in the language of origin) would help users of the planetary nomenclature who wish not only to write but also talk about planetary science, this is why the Commission on Planetary Cartography of ICA established a pronunciation guide, independently from IAU (Hargitai and Kereszturi 2010).

The UNGEGN defines a geographical name as a name applied to a feature on Earth (UN 2006), therefore is not concerned with extraterrestrial names.

National names authorities have also no responsibility for the creation or approval of a standardized national variant of the planetary gazetteer since planetary surfaces do not belong to any national territories, they are *terrae nullius:* “Outer space, including the moon and other celestial bodies, is not subject to national appropriation by claim of sovereignty, by means of use or occupation, or by any other means” (UN 1966).

A planetary feature name nor is an exonym—a “name used in a specific language for a geographical feature situated outside the area where that language has official status, and differing in its form from the name used in the official language or languages of the area where the geographical feature is situated” (UN 2002)—neither an endonym—a name from within a social group (Jordan 2010) or officially: “name of a geographical feature in one of the languages occurring in that area where the feature is situated” (UN 2002). This is similar situation to undersea features (Maciej 2010).

This *terra nullius* situation makes it hard to find the appropriate national or international body which may be responsible for developing localized variants.

Consequently, localizations (generally: translations)—which is practiced day by day—are created by journalists, cartographers, popular science authors, online editors of various magazines and websites, all on an *ad hoc* basis. Today efforts to create standardized language variants of the Planetary Gazetteer are made in Russia, Hungary, China, and probably in Japan and South Korea as well (IPCD 2011). If space permits (or the map editor decides to do it so), transformed variants may appear next to the official, making these maps bilingual and in many cases, biscriptual. Other countries use the official forms or *ad hoc*—or, in the case of large Lunar albedo features, traditional—translations.

The ultimate goal of this work was to confirm or disprove the positive effect of our efforts on the localization of planetary nomenclature; using these results we can move forward on developing more localized gazetteers or alternatively we can seek for other methods of making planetary maps more comprehensive for non expert map users.

## 9.3 Methods

In this paper results of two surveys are reported. In both surveys the questions were qualitative, while Kereszturi (2009) have conducted a survey in which quantitative estimates had to be made by students, related to Martian environments from an astrobiological point of view.

In 2007, a questionnaire was distributed to various age groups, asking them to characterize Mars. In 2011, a more focused survey was made in four middle school classes in Hungary(Kürt Alapítványi Gimnázium, Budapest, 10a, 10b, 11a, 11b classes) (two classes studying geography in English, two in Hungarian, 16–17 year old pupils). They were given a questionnaire and maps. They first had to give a short description of Mars without a map—this task was similar to the 2007 survey. Then the students got one map-specific task. They got two sets of maps of Mars. Half of them got a map showing a nomenclature in the official form (in Latin), the others got the standardized Hungarian language variant (Hargitai et al. 2010; Hargitai and Kereszturi 2002). The base map and layout of these variants were similar. After receiving the maps, they again had to describe Mars, but now using these medium and small scale maps (Figs. 9.1, 9.2). The medium scale maps were cutouts from the global topographic map of Mars (Hargitai 2008q).

It would have been interesting to include P. Lowell's so-called *Lowellian canal network nomenclature* which probably would result in a completely different

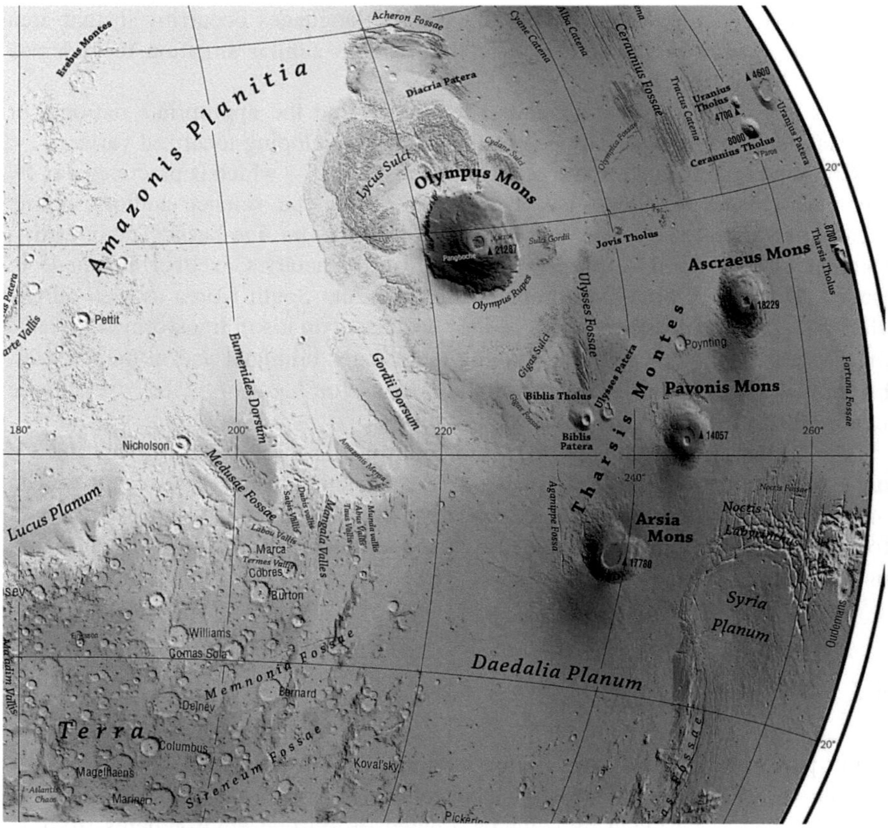


**Fig. 9.1** Detail of the map of Mars, with official IAU (Latin) nomenclature (all maps by author)

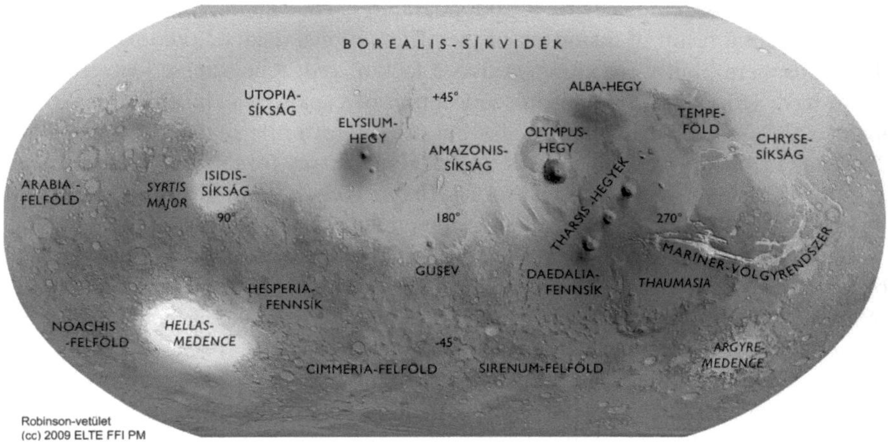


**Fig. 9.2** Small scale map of Mars showing the Hungarian nomenclature

picture of Mars (Sagan and Fox 1975; Lowell 1908); but this system of place-names would not fit the topographic map of Mars.

The survey was made using the map of Mars, and not of the Moon, with which the students are more familiar (at least visually), because the Lunar nomenclature uses traditionally false descriptor terms (seas, lakes, etc.) (Whitaker 1999; Greeley and Batson 2007), which would unnecessarily add one more variable which affects the map reader's concept of the surface.

## 9.4 Results and Discussion

### *9.4.1 Concepts of Mars Without a Map*

What is the surface of Mars like? Two independent surveys have been made: the first asked this question from 3 to 66 year old persons, including university students. The results of this survey were reported in detail by Hargitai (2008b). In the following the results concerning only Mars are presented.

"Wasteland" was a common answer in all age groups. Even university students responded that they previously imagined Mars as a global plain. This is in good agreement with the nineteenth century picture of Mars in which Mars was seen as a global plain (with some possible vegetation). Topographic features were never seen prior to spacecraft observations from near orbit, because from the Earth it is not possible to see shadows on Mars: Mars, seen from the direction of the Sun, is always in or near full phase. (We can easily observe this phenomenon on the Moon with a small telescope: during full Moon phase, topographic features can't be seen, only albedo markings divide the surface, while near the terminator line, crater rims cast long shadows, emphasizing local topography). The lack of visible topographic features were a result of distortion of the observational data set (and partly due to the atmosphere), but were interpreted as a lack of *real* topographic features. This could have led to the common misconception of canals, which, according to the theory, are used for irrigating the dry but otherwise fertile plains (Lowell 1908). This way the theory of a flat Mars helped paving the way for the theory of the canal system of Mars. It is also easiest to imagine a global plain, since it requires no imagination of more complicated geographic features.

Between the age of 12 and 18 the color red was often mentioned, but this observational fact, known by anyone who have ever seen Mars in the night sky, was interpreted differently: a 12-year-old explained the reddish color as “lava”, some 15–22-year-old physical science students said that it is because of its iron content, while one student of humanities attributed the color to the atmosphere of Mars. This latter explanation coincides with the history of discovery of Titan, which orange color was first interpreted incorrectly as the color of its surface by G. Kuiper (1944).

A 15-year-old “terraformed” Mars: he placed “mountain chains, valleys and volcanic mountains” onto Mars. Apart from university students in Earth Sciences, only one respondent mentioned craters and the polar caps. The other answers mentioned volcanoes, river valleys, dust and sand.

This survey showed that several different views exits about Mars, and these images are dependent on the age of the respondents.

### *9.4.2 Concepts of Mars Before and After Viewing the Maps*

The next survey was made in 2011 in four middle school classes where students could also study the maps. Although the students got the same maps, their responses were not at all uniform in any of the groups (Fig. 9.3).Most students in this survey mentioned mountains, craters and hills the most, plains only ranked fourth place. This is in contrast with a previous study in which plains was the most common answer. Many students gave more details: rocky surface, res soil, dust, buttes, snow caps, or mentioned other aspects of its surface: dry and past water (Fig. 9.3).

### *9.4.3 Latin Names*

Students reading the map in *Latin*, used the Latin names for surface description, often without any comment. Names with *Rupes, Sulci, Planitia* could not be interpreted by any of the respondents, they only “copied” the names. Valles Marineris was interpreted as ravine or long valley.

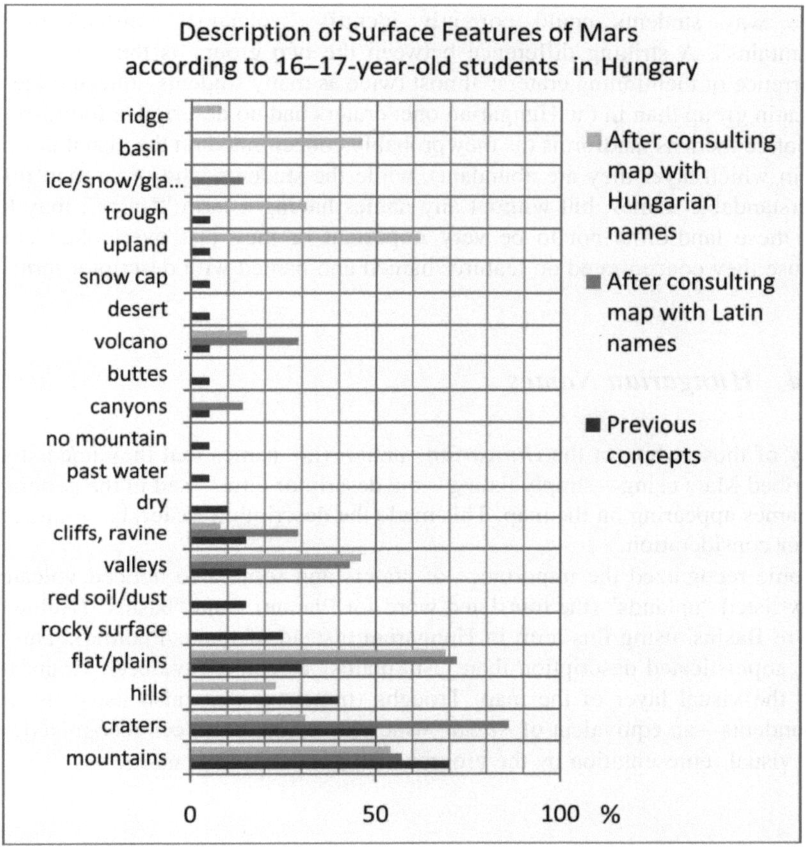


Fig. 9.3 Results of the survey

Interestingly, valleys were equally mentioned by both groups. Probably users of the Latin version could easily identify them from their visual appearance together with the resemblance of the Latin word *Vallis* to the English Valley.

*Plana* were understood as plains, *Mons* and *Paterae* were described as high mountains, Tharsis Montes as mountain range. Craters were the most frequently mentioned as typical Martian landscape features.

The group who worked with the Latin names could not use the descriptor term parts of place names as handles so they had to figure out the geographical nature of the features. They had to use the map's visual tools: shapes and shaded relief representation of landforms. So they have to invent English (or Hungarian) words instead of the Latin ones, in which they performed well. In some way, even better than those who got the Hungarian version, transparent in meaning: descriptor terms reflect the shape of features (morphology), while looking at the shape of them, their origin/genesis (or geological nature) may have also become clear for the students.

These way students could correctly identify "volcanoes" instead of just "mountains". A striking difference between the two groups is the frequency of occurrence of mentioning craters: almost twice as many students noticed craters in the Latin group than in the Hungarian one: craters had no descriptive term, so they did notice them as landforms (as they probably concentrated on the visual information in which layer they are abundant), while the students who got a map full of understandable names, but without any names having a term "crater", may have seen these landforms not to be very important or they just overlooked craters because they concentrated on features named and named with descriptor terms.

### *9.4.4 Hungarian Names*

Many of those who got the *Hungarian version* (i.e. names that they understood), described Mars using—simply listing—the descriptor terms used in the geographical names appearing on the map. This made the description easier, but required no deeper consideration.

Some recognized the importance of craters and some also noticed volcanoes. Many listed "uplands" (the translated word for Planum), and "basins" (Hellas and Argyre Basins, using this term in Hungarian instead of Plains/Planitia). This is a more sophisticated description than just "plains" as could have been "translated" from the visual layer of the map. Troughs (trenches) were also listed by many respondents—an equivalent of *Fossae* which have not at all been recognized from their visual representation by the group which got the Latin names.

## 9.5 Discussion

Of the landscape types described before consulting the maps, some concepts have disappeared completely from the later description: deserts, buttes, soil and dust, rocky surface are not mentioned any more. Buttes are too small to appear on a map, surface characterization is not the task of a topographic map. Arid and dry nature of the surface can't be read from a topographic map. *These very important concepts which described the nature of Martian landscape correctly, have been exterminated and overridden by the stronger effect of the nomenclature of the map.* However, it is very important that these factors remain in the horizon of the student's knowledge on Mars: maps—especially computer generated automated maps which are not suitable for showing features smaller than what is visible at a particular scale—should have solution of communicating the surface texture and the existence of smaller landforms in some way.

*The language of nomenclature does change the perception of the surface features of a planetary body for a general map reader.* It changes preconceptions, eliminating some—even important and true—concepts, and makes others even stronger. If the meaning of a descriptor term is opaque (used in a language not known by the map reader) (Kadmon 2000), they had to interpret the features themselves; this gives more importance to the visual representation of features. If the meaning of a descriptor term is known, it makes it easier to interpret features, but since it is an easy task, the map reader does not have to rely on visual representation and will less notice features shown but not named. Transparent names help interpreting features whose nature is hard to decipher from its visual representation.

## 9.6 Conclusion and Future Directions

Planetary maps are representations of the physical reality but since the cartographers' representation methods are taking aesthetic and other human factors into consideration—such as classification, generalization and naming landforms (selected land areas)—are subjects of cognitive reception therefore their interpretation by non experts depend on the used visualization and naming methods. This work showed a survey which was focusing on one variable: the language of the nomenclature—or, more precisely, the language of descriptor terms. Hungarian (in this case, transparent) and Latin (opaque) terms were shown to a non expert target audience.

The survey showed that cartography is indeed a form of art, since just like literary texts, it is highly dependent on the interpretation and preconception (or, previous knowledge) of its reader.

The respondents who consulted a planetary map gave a more detailed description of Mars, which was in accordance of the first hypothesis of this work ("first consulting a planetary map [. . .] will provide a basis for a more detailed cognitive map than all previous knowledge on that planet") but in these new concepts they ignored several parts of their previous description. It shows that the first encounter with a planetary map caused a kind of paradigm shift in which they abandoned large parts of their previous view of the surface, at least, for this description task.

The second hypothesis was that "if the opaque descriptor terms [. . .] are translated to a language transparent to its reader, the map reader [. . .] will have a more realistic (cognitive) view of the surface [. . .]". This work showed that the situation in not this simple. Both methods—having an opaque or a transparent descriptor term—have their advantages and disadvantages, which should be taken into account when the visual representation methods and additional data or texts that appear on the map are planned and realized.

Translation of the opaque (not understandable) nomenclature gives a different view which is neither better nor worst than the Latin version. This conclusion disproves the need for the translation of the Latin names in order to achieve a better interpretation of a map showing an extraterrestrial surface. However, it sheds light to those elements which need to be emphasized by other methods. This is true for both cases. In the case of those cultures and languages, which use a writing system other than the Roman script, transformation of the script (together with translation) may still be needed. In the maps using translated names, more visual or other emphasis should be given to identify features without descriptor terms. If the original Latin nomenclature is used, more emphasis should be given to the explanation of the nature of the fully named features. In both cases, the true nature of features with no terrestrial equivalents should be explained together with those features and characteristics of the surface that can't have names because they are not individual features but commonly found or are common properties of the whole surface. It is a rule that "official names will not be given to features whose longest dimensions are less than 100 m" (USGS 2011). But such features may give a basic characteristic of a particular planetary surface.

The goal of a cartographer is therefore to make the following types of information noticeable even for the non expert map reader:

(a) Previous, correct concepts not displayed on a general map (desert, rocky, buttes, soil, dust)
(b) Characteristics that are lost because of the name: craters (not noticed because they have no generic), volcanoes (named "mountains"), *fossae* (nor the descriptor term, nor the visual representation helps it interpretation as tectonic graben), etc., and
(c) Other, planet-specific special information, in the case of Mars, the differentiation of upland versus lowland versus basin (which are also plains at the same time).

Amongst the possible solutions are the use individual symbols, textures, marginal text or in-map annotations. In the case of maps for the general public, these elements could be designed individually for a specific map and do not necessarily have to—but may—follow the standardized representational methods used in planetary geology or morphology (Nass et al. 2010).

The set of colors used for planetary topographic maps also affects the reader's concept of the surface. We have used white to brown color scheme for Mars, to avoid bluish hues that may be associated with water bodies. For the same reason, Lazarev and Rodionova (2011) developed a color scale for Venus in which they used purple to represent areas below 0 height level (mostly large basins).

A consensual solution to this problem may be to include both nomenclature variants together with additional symbols and explanatory texts. In the virtual reality we have the option of turning off the nomenclature layers which would provide a view without any "diverting" textual description, interpretation or preconception. This variable view can be realized using online maps.

It should also be noted, that independently from the results of this research, the standardization (or, alternatively, a complete elimination) of the already existing locally used language variants of the planetary nomenclature remains a problem awaiting solution.

## References

Greeley R, Batson RM (2007) Planetary mapping. Cambridge University Press, Cambridge/New York

Hargitai HI (2006) Planetary maps: visualization and nomenclature. Cartographica 41 (2):149–164. doi:10.3138/9862-21JU-4021-72M3

Hargitai H (ed) (2008a) Topographic map of Mars, Lambert projection. http://planetologia.elte.hu/1cikkeke.phtml?cim=planterkepeke.html. Accessed 15 July 2011, KAVÜCS, Budapest

Hargitai H (2008b) Csillagászati ismeretek és téveszmék óvodás kortól idős korig, [Astronomical facts and fallacies from pre-schoolers to elderly people, in Hungarian]. Új Pedagógiai Szemle 58:122–148

Hargitai H, Kereszturi A (2002) Javaslat magyar bolygótudományi szaknyelvi norma létrehozására [Suggestions for a Hungarian language standardized planetary nomenclature and terminology, in Hungarian]. Geodézia és Kartográfia 2002:26–32

Hargitai H, Kereszturi Á (2010) Towards the development of supplements to the gazetteer of planetary nomenclature. In: Proceedings of EPSC2010-865. European planetary science congress 2010, Rome, 19–24 Sept 2010

Hargitai H, Császár G, BércziSz KÁ (2008) Földön kívüli égitestek geológiai és rétegtani tagolása és nevezéktana [Geological and stratigraphical units and the nomenclature of extraterrestrial planetary bodies, in Hungarian]. Földtani Közlöny 138:323–338

Hargitai H, Kozma J, Kereszturi Á, Sz B, Dutkó A, Illés E, Karátson D, Sik A (2010) Javaslat a planetológiai nevezéktan magyar rendszerére. [Suggestions for a Hungarian language system of planetary nomenclature Hungarian]. In: Meteor Csillagászati Évkönyv [Meteor Astronomical yearbook]. MCSE, Budapest, pp 280–302

IAU (1970) Transactions of the International Astronomical Union, in Menzel, D.H., Minnaert, M., Levin, Boris, and Dollfus, Audouin, 1971, report on lunar nomenclature. Space Sci Rev 12:136–186

Ingarden R (1974) The literary work of art. Northwestern University Press, Evanston

IPCD 2011: international planetary cartography database. ICA commission on planetary cartography//Eötvös Loránd University, Budapest. http://planetologia.elte.hu/ipcd/. Accessed 15 July 2011

Jordan P (2010) The endonym—name from within a social group. In: Proceedings of trends in exonym use proceedings of the 10th UNGEGN working group on exonyms meeting, Tainach, 28–30 Apr 2010

Jordan P (2011) Trends of exonym use in European school atlases. In: Proceedings of international cartographic conference, Paris, No. CO-376, 7 July 2011

Kadmon N (2000) Toponymy: the lore, laws and language of geographical names. Vantage Press, New York

Kereszturi Á (2009) Visualization in the education of astrobiology. In: Hegedűs S, Csonka J (eds) Astrobiology: physical origin, biological evolution and spatial distribution. Nova Publishers, New York, pp 131–141

Krygier JB (1995) Cartography as an art and a science? Cartogr J 32:3–10

Kuiper GP (1944) Titan: a satellite with an atmosphere. Astrophys J 100:378–383

Lazarev E, Rodionova J (2011) Venus mapping at small scale: source data processing and cartographic interpretation. In: Ruas A (ed), Advances in cartography and GIScience. doi 10.1007/978-3-642-19214-2_17

Lowell P (1908) Mars as the abode of life. Macmillan, New York

Maciej Z (2010) Polish geographical names of undersea and Antarctic features. The names which "escape" the definitions of exonym and endonym. In: Trends in exonym use proceedings of the 10th UNGEGN working group on exonyms meeting, Tainach, 28–30 Apr 2010

Nass A, van Gasselt S, Jaumann R, Asche H (2010) Implementation of cartographic symbols for planetary mapping in geographic information systems. Planet Space Sci 59:1255. doi:10.1016/j.pss.2010.08.022

Ong WJ (1982/2002) Orality and literacy: the technologizing of the word. Routledge, London/ New York

Sagan C, Fox P (1975) The canals of mars: an assessment after mariner 9. Icarus 25:602–612

Searle J (1997) The construction of social reality. Free Press, New York

Shingareva KB, Zimbelman J, Buchroithner MF, Hargitai HI (2005) The realization of ICA commission projects on planetary cartography. Cartographica 40:105–114

UN (1966) UN 2222 (XXI). Treaty on principles governing the activities of states in the exploration and use of outer space, including the moon and other celestial bodies. http://www.unoosa.org/oosa/en/SpaceLaw/gares/html/gares_21_2222.html Accessed 15 July 2011

UN (2002) Glossary of terms for the standardization of geographical names: United Nations group of experts on geographical names. Department of Economic and Social Affairs Statistics Division, New York

UN (2006) Manual for the national standardization of geographical names: United Nations group of experts on geographical names. UN Department of Economic and Social Affairs Statistics Division, New York

USGS (2011) Gazetteer of planetary nomenclature. International astronomical union working group for planetary system nomenclature. http://planetarynames.wr.usgs.gov/Page/Rules Accessed 15 July 2011

Whitaker EA (1999) Mapping and naming the moon: a history of lunar cartography and nomenclature. Cambridge University Press, Cambridge

Wittgenstein L (1922) Tractatus logico-philosophicus. Routledge and Kegan Paul, London

# Chapter 10
# Internet Mapping Education: Curriculum Technology and Creativity

**Rex G. Cammack**

**Abstract** Educators are always looking to update and more effective teach their students. In the area of Internet Mapping the changes in technology can quickly isolate course materials as obsolete. This research uses a case study approach to looking at the issue of curriculum and technology in the planning and delivery of a course that both education and empower creativity in students. The process of developing educational goals and assessing how to implement them are two critical steps in the curriculum process. The results of this step formalized the process of reviewing and selecting technologies for Web Mapping. At the end of this process the case study selected Adobe Flex and its IDE Flash builder as its development tool. With the technology selected the next process was to design exercised for student that develop there conceptual and technology skills in the area of Web Mapping. The overall goal of this project was to enable students to go beyond their learning task knowledge to foster link between technology and creativity. The conclusions drawn from this case study are that a balance between technical capabilities and creative possibility exist when using the Adobe FLEX and its IDE Flash builder. The environment allows students to utilize both their skills in application development and graphic design to create complex interactive Web Mapping products.

## 10.1 Introduction

In the process of educating students in the area of Internet Mapping, faculty must choose systems and tools that work in tandem with their educational goals. The number of options for these types of education technologies is almost unlimited. How does one choose and environment that will provide education flexibility and

R.G. Cammack (✉)
University of Nebraska Omaha, Omaha, USA
e-mail: rcammack@unomaha.edu

L. Zentai and J.R. Nunez (eds.), *Maps for the Future*,
Lecture Notes in Geoinformation and Cartography 5,
DOI 10.1007/978-3-642-19522-8_10, © Springer-Verlag Berlin Heidelberg 2012

comprehensiveness without having students change for one technology to another for each project? There are numerous whys to approach this question in this case study we will goes through the tasks of setting education goals, implementation assessment, technology assessments, exercise plan and creativity possibilities. This approach is not deterministic in the sense that there is only one possible solution or even a solution that meets all the criteria. One of the goals for this research is to just simple identify most of the specific issues in regard to educating students in the area of Internet Mapping in the current environment. A second goal of the research its to look at the education technology in two ways. First is the whether the technology can provide a learning environment for specific learning generic learning goals, while the second part examines the ability of the technology to foster creative ideas.

## 10.2 Education Goals

In the university education environment the term curriculum development and management is used. For many faculty member and program groups this term concept is only managerial task that must be done. Looking deeper in to this practice one learns that there are many facets to examine and consider. There are three-principle parts to curriculum to consider within the university environment.

- University Education Outcomes
- Programs Outcomes
- Course Outcomes

Most have the time faculty and teachers look solely at the later of these outcomes because it's the one that have direct control of over the time of the course. Faculty members develop plan of study for the course that meets the needs of the course with little to no consideration of the other learning outcomes that the university and program need. The bottom up appropriate has been employed widely throughout university and has served well for the most part. One of the reasons for this is that faculty members and course content can change rapidly. In the United States of America (USA) university systems most courses have a curriculum structure found a university catalog (University of Nebraska Omaha, 2010). These consist of a title, number of credit hours, course description around 80 words in length and prerequisites for the course. These course catalog entries can be used for years with no changes or updates. As an effective curriculum tool, it is evident that there not all that effective yet this method is the primary direct contact with perspective students.

In this age of higher education there is a need to go beyond the current state of education delivery and focus on new approach. One of the approaches changes the conception of curriculum from static to a "living curriculum" (Bath et al. 2004). In this context curriculum is view and vibrating and changing much like the changing world we live in. Bath et al. (2004) examine living curriculum in the Australian higher educational system. Not unlike the USA higher educational system there is a debate on how to education students for an ever-changing world. The crux of this debate is focused on program knowledge verses generic skills. University as a whole identify that the principle task is education student to communicate, problem solve, analysis, integrate and creatively think. The argument is on how to achieve this. One approach is to focus on this skill specifically and have all students master these generic skills in assignment courses (Harvey 1993; Gash and Reardon 1998; De La Harpe and Radloff 2000). Others have argued that only through effective disciplinary education do students learn and apply these generic skills (Boyatzis et al. 1995; Kemp and Seagraves 1995; Diamond 1998, and Drury and Taylor 1999). This two extreme in curriculum approaches have been bridged but neither side seems to like the option (Gash and Reardon 1998). Ideas like tacked-on, afterthought bolt-on (Bowden et al. 2000; Pearson and Brew 2002) and check list are seen as failing to delivery acceptable educational outcomes.

The later approach does have appeal at both at the program and course outcomes level. In a closely related context the efforts by DiBiase et al. (2007) to review and establish the merits of the Geographic Information Science and Technology Body of Knowledge (GIS&T BoK) (AAG, 2006) shows a exhaustive set of disciplinary skills that students should learn and master. These skills are closely related to Web Mapping and one might think that all of web mapping is covered in the GIS&T BOK. But as Dibiase et al. (2007, p. 119) say "that disruptive new technologies, innovative sciences, imaginative applications, and the dynamics of human and physical landscapes that give rise to the GIS&T field in the first palace, will continue to drive changes."

In the field of Information Systems Gorgone et al. (2003) developed a comprehensive curriculum of undergraduate programs. The curriculum like the GIS&T BoK was a discipline outcome focused approach. This middle ground approach to curriculum development focuses on assessment of market demands for graduate. This special consideration for the need of the employment market does focus a program to continually update is educational system. However one can argue that this approach will only develop time sensitive vocational skills and graduates will be ill prepared to handle changes in the work place.

Designing a course focused on Web Mapping technologies how does the faculty member approach this education. Web Mapping requires some very specific skill sets. By choosing a development environment vocational training looms and universities are quick to say they are not vocational by nature in the USA. One approach is to use a wide variety of development environments. In this environment student are forced to learn not a system but a method of problem solving. Hear the student must create a knowledge set that says how do I do this in this new system. An example would be how do I declare a variable type. In every programming language declaring variable is a basic concept. So by switching environment multiple times within the course student will develop a method for learn the next system. This problem-solving outcome fits nicely in all three learning outcomes. Yet this outcome result is extremely difficult to achieve. It takes a long time do develop this knowledge even for professional in the field for decades and many times professionals really never develop this skill.

At the end of this process for developing educational goal a pragmatic approach was selected. First a comprehensive look at university education outcomes was completed. The following outcomes were identified.

- Critical Thinking
- Problem Solving
- Communication
- Diversity
- Cooperative Learning
- Team Building
- Social Values
- Creative Thinking and Expression

The second phase of learning outcomes looked at degree program goals. In this case study the University of Nebraska Omaha degree program in Geography was considered. This program consists of a Bachelors' of Arts and Bachelors' of Science program. The course was design to meet the needs of both disciplinary programs. These two programs are similar in education outcomes goals but with variances in important of the goals:

- Understanding of Physical Process
- Understanding of Humans behavior and Societal Variances
- Relationship between Environment and Humans
- Analysis of Spatial Events
- Environmental/Human Landscapes
- Resources and Human interaction
- Geospatial Representation and Analysis
- Regional Difference
- Social Acceptance
- Qualitative and Quantitative spatial Methods

For the course outcome there will be discussed in more detail later in the paper.

### *10.2.1 Implementation Assessment*

The process of implementation assessment is difficult to do. At the level of university education outcomes most choose a post hoc approach by looking at the success of students graduating for the universities programs. Issues like profession certification and accreditation are procedural methods of assessment. Additional universities will interview students near graduation to determine whether they are successfully finding employment. These methods are helpful but most of the time does little to assess the generic skills for students. Bath et al. (2004) employed a survey approach that focused on students after a major change to a degree program was done for a music program at the University of Queensland. The focus of the research was to examine the effectiveness of doing a complete review and change to music school programs that specifically combine the learning outcome for the university and the discipline within each course. The results of the research showed that students from the updated program were doing better than other students at the university.

In this project a different approach was taken. A single class was used. The course was an advance Geographic Information Science course. The course outcomes where reassessed and more emphasis was placed on incorporating aspect of university and degree program goals. Some of the changes to the course were.

- Team Projects
- Alternative Demonstration of work i.e. Poster Presentation
- Quantitative analysis
- Visualization (Conceptual and Numerical)
- Comparative Solutions
- Creative and Limited Guided Task
- Independent Problems
- Solution Integration
- Critical Self Evaluations

Many times the new tasks where assigned in a way that they covered several generic skills. Also students did new assignments that were unique combination of the learning outcomes. The goal was to have students doing these learning outcomes several times without directly focusing on them. The method of integrating several learning outcome and repeating of reinforcement was questioned but this was the approach settled.

## *10.2.2 Technology Assessment*

In this part of the project specific discipline skills were considered. The Advanced GIS course was reorganized around the core concept of Geographic Information Services. Nyerges and Jankowski (2010) explain that Geographic Information can be explained as a system, science or service. In previous version of the course aspect of Advance systems and science were the focus but change in GIS in the last 5 years have led to the development a professional need for better understanding of information services. Within the new redesign the focus was on these GIS skills.

- Spatial Data
- Spatial Data Services
- Implementation Issues for Services
- Topic Specific Geographic Services
- Service Design
- Service Deployment
- Services on a Network
- Services on the Internet

- Service Use
- Standard Clients for Service Use
- Custom Clients for Service Use
- Multiple Spatial Data Services
- End User Clients

The approach here was to selecting a combination of Web Mapping technology and development environments. Students would use these technologies and develop projects to showcase their understanding to these disciplinary skills. A second consideration was to have student use technologies that are common with in the USA geospatial professional workplace. The basic design of Web Mapping services is showing in Fig. 10.1. The basic element of a GI service starts with a network most commonly the Internet. A server on the network host a GI service and clients or other service consume the data for the GI Services. Within the server any number of designs can be deployed but the simple structure is that GI server is hosted directly or through a Web servers. The GI server will with store within of get form external devices the spatial data and/or spatial processing functionalities.

In the course issues like servers, network, web servers, database service, and middleware were explained but the students did not work with them. The main focus for the students was on the build of the GI services and clients.

The GI server software used in the course was ArcGIS server 10.1 deployed on Microsoft Internet Information Web server with the DotNET framework. Student where asked to design both services and Web application using the ArcGIS Server application. For the more detailed client development for specific Internet Mapping applications students used the combination of Adobe FLEX 4 and Adobe Flash builder 4. This development environment was selected for several reasons:

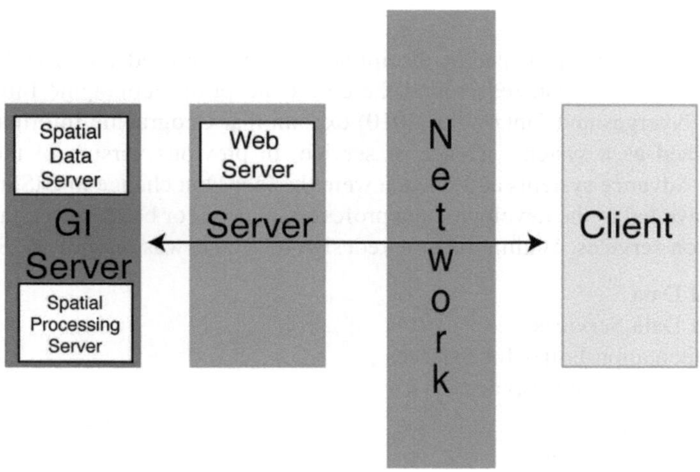


**Fig. 10.1** Conceptual structure of web mapping environment

- Integrated Development Environment (IDE)
- FLEX Framework Functionality
- API for Google and ArcGIS Server
- Skinablility
- Modularity
- Multiple Platform capabilities

Client development is the fastest growing part of Internet Mapping. Mostly this is due to the availability of large GI servers open to anyone. Google GI servers host several different worldwide mapping databases such as Google Maps and Google Earth (Google 2011). With this type of service available numerous mapping clients are develop either for wide user or focused users. One of the many aspects of developing client Internet Mapping application is the deployment environment. One of the major advantages for Adobe Flex environment is that is build both application for the computer desktop and web application that run with in a web browser. These are two of the three main client deployment environments the third environment is the mobile environments such as Android iOS. The mobile environment is the most complex and varied of this three. Mobile computer is the newest of the three, so there is still a large amount corporate maneuvering at time of this research. In the future (such as the next time the course is taught) a reviewed implementation will need to be considered.

### 10.2.3 Exercise Plan

All the exercises in the course were design to consider all three learning outcome discussed so for with in the research. This list of exercises for the course show the wide difference in task student did with in the course. At first glance the effects of the university and program outcomes are not apparent.

- Multi-layered Environmental GI service
- Multi-Server Demographic GI service*
- Desktop Client for GI service
- Desktop Client for Multiple GI services*
- GI Service Standardization for Integrated Internet Mapping Website*
- Topic Based Client Web Application*
- Interactive Client Applicants
- Desktop Client Application for Corporate Task*
- Final Project Student Focused GI services and Client Application

*group assignments

The course was dividing into 5-h blocks each week of the 15-week semester. The first hour devoted to lecture on core ideas and concepts. A second hour was developed technology issues regarding GI services and management. The third hour would focus on the client development environment and the final 2-h work laboratory time for student to work on their assignments.

The each assignments students developed new disciplinary skills that would use in following assignments. Skills in both GI services and Client development progress until the final project where student had created based on their own ideas. This finals project is consider and demonstration of knowledge and could be used for a detail analysis for the learning outcomes at the university, program and course. The topic of the assignment focused on program outcomes so as students worked to develop their systems they had to understand and consider broader program issues.

### *10.2.4 Creative Possibilities*

The idea of creativity for the most part in higher education is identified with the fine and liberal arts disciplines of the university. Yet in Geography creative expression with maps has long been understood. This creativity has led to some great maps and also some bad ones but visual representation as in writing is not just a static concept but has a sense of expression that cannot be over looked. As each assign was defined to students few and few detail of the final look of the work was defined and students were asked to think not only of function but also form or their project. Design was given value with in the assessment of projects and student with creative solution were given opportunities to show them to the rest of the class and explain their ideas behind their creative solutions.

## 10.3 Conclusions

The process of assessing and re-design a university course now long can be done in the vacuum of an individual faculty member interest and desire. Universities are more and more integrating student learning across the programs and courses students take. The model of learning outcomes will have a greater influence on course structures and activities. Currently the model of learning outcome suggest that the best outcomes are achieve by excellent course content education that looks at considers within it delivery issues of program and university outcomes. In this case study an Advance GIS course was re-design with special consideration made to outcomes for both the university and program. The selection of education technologies led to use of robust put a limited number of technologies. Students were asked to repeat aspect of GI services several times during length of the course to re enforces concepts and to consider closely the effect of their decision on the overall technology implementation and societal importance.

## References

Bath D, Smith C, Stein S, Swann R (2004) Beyond mapping and embedding graduate attributes: bringing together quality assurance and action learning to create a validated and living curriculum. High Educ Res Dev 23:313–328

Bowden J, Hart G, King B, Trigwell K, Watts O (2000) Generic capabilities of ATM university graduates: final report to DETYA. Melbourne Australian technologies network teaching and learning committee

Boyatzis RE, Cown S, Kolb D (1995) Innovation in profession education. Jossey Bass, San Francisco

De La Harpe, Radloff A (2000) Supporting generic skill development: reflections on providing professional development for academic staff. In: Lifelong Learning Conference: selected papers from the inaugural international Lifelong Learning Conference. Rockhampton, Central Queensland University, pp 41–47

Diamond R (1998) Designing and assessing courses and curricula. Jossey Bass, San Francisco

Dibiase D, DeMers M, Johnson A, Kemp K, Luck AT, Plewe B, Wentz E (2007) Introducing the first edition of Geographic Information Science and Technology Body of Knowledge. Cartogr Geogr Inf Sci 34:114–120

Drury H, Taylor C (1999) Providing the discipline context for skills development: report on the production on an interactive video for oral and visual communication in the biological sciences. Paper presented at the HERDSA annual international conference, Melbourne

Gash S, Reardon DF (1998) Personal transferable skills for the modern information professional. J Inf Sci 14:285–292

Google (2011) Google maps. http://maps.google.com

Gorgone J, Davis GB, Valacich JS, Topi H, Feinstein DL (2003) IS 2002 model curriculum and guidelines for undergraduate degree programs in information systems. Commun Assoc Inf Syst 11:11–233

Harvey L (1993) Quality assessment in higher education: the collected papers of the QHE project. University of Central England: Quality in Higher Education, Birmingham

Kemp IJ, Seagraves L (1995) Transferable skills can higher education deliver? Stud High Edu 20:315–328

Nyerges TL, Jankowski P (2010) Regional and urban GIS: a decision support approach. Guilford, New York

Pearson M, Brew A (2002) Research training and supervision development. Stud High Edu 27:135–150

# Chapter 11
# Spatial Knowledge Acquisition in the Context of GPS-Based Pedestrian Navigation

**Haosheng Huang, Manuela Schmidt, and Georg Gartner**

**Abstract** GPS-based pedestrian navigation systems have become more and more popular in recent years. This paper reports a work in progress on investigating the differences of spatial knowledge acquisition with different interface technologies in the context of GPS-based pedestrian navigation. The acquisition of spatial knowledge based on mobile maps, augmented reality, and voice is analysed and compared in a field test in the city centre of Salzburg (Austria). This paper presents the methodology and interprets the results. The results raise some hints for future mobile navigation system development, which might need to consider not only how to effectively assist users' navigation tasks, but also how these systems affect users' spatial knowledge acquisition.

## 11.1 Introduction

People in unfamiliar environments often need assistance to reach a specific destination. Mobile pedestrian navigation systems are designed for this purpose. In order to facilitate pedestrians' navigation tasks, navigation systems need to effectively communicate/convey route information (directions) to pedestrians.

Map is an important interface technology when communicating route information. It can help users to get an overview of an area. Radoczky (2004) shows that maps, even presented on mobile devices with small screen, are the most efficient tool for describing route directions. Voice-based guidance is also a useful tool for navigation. In additional to metric-based instructions which are often used in car navigation systems, semantic-based instructions enriched with landmark

H. Huang (✉) • M. Schmidt • G. Gartner
Institute of Geoinformation and Cartography, Vienna University of Technology, Vienna, Austria
e-mail: haosheng.huang@tuwien.ac.at; manuela.schmidt@tuwien.ac.at; georg.gartner@tuwien.ac.at

L. Zentai and J.R. Nunez (eds.), *Maps for the Future*,
Lecture Notes in Geoinformation and Cartography 5,
DOI 10.1007/978-3-642-19522-8_11, 

information are also proposed for route communication (Rehrl et al. 2010). Recently, mobile augmented reality (AR), which enhances the real world camera view with virtual information overlays, is another promising approach for conveying route information. Walther-Franks (2007) shows that AR is very suitable for navigation as it puts route instructions directly into the real visual context of a user.

Spatial knowledge acquisition is needed to build mental representations that are essential for wayfinding and other spatial tasks. With sufficient spatial knowledge about an environment, people can still find their way when navigation systems fail. Currently, more and more people are relying (or even over-relying) on mobile navigation systems. Therefore, in additional to the effectiveness in supporting wayfinding, it is also very important to investigate how these systems affect the acquisition of spatial knowledge.

There is some research focusing on empirically studying the acquisition of spatial knowledge in the context of pedestrian navigation. Gartner and Hiller (2009) investigated maps with different display sizes, and show that display size influences spatial knowledge acquisition during navigation. Ortag (2005) studied the differences of spatial knowledge acquisition with maps and voice when guiding wayfinders. Krüger et al. (2004) compared the impact of different modalities (i.e., audio and graphics (specially, images indicating route directions)) on spatial knowledge acquisition during navigating in a zoo, and conclude that the acquisition of route knowledge is much better than that of survey knowledge. In Aslan et al. (2006), the differences in acquiring spatial knowledge with and without technology (e.g., mobile maps versus paper maps) were studied. It is important to note that most of the above studies employed the "Wizard of Oz" prototyping (Wikipedia 2011) (e.g., without using the GPS). In contrast, Ishikawa et al. (2008) compared the acquisition of spatial knowledge with GPS-based systems, paper maps and direct experience of routes, and show a poorer performance of subjects using GPS-based system. However, to the best of our knowledge, none of the field test compares the influence of mobile maps, AR, and voice on spatial knowledge acquisition in the context of GPS-based pedestrian navigation.

This paper presents an on-going work on empirically studying the differences in spatial knowledge acquisition with different interface technologies, comparing mobile maps, AR, and voice in the context of GPS-based pedestrian navigation. This research is part of the ways2navigate project, which is a project of Vienna University of Technology, Salzburg Research, FACTUM, TraffiCon and WalkSpace Mobilität. It aims to investigate the suitability of voice-based and AR-based interface technologies in comparison to mobile maps for conveying navigation and route information to pedestrians. Two iterative field tests are planned in the ways2navigate project. For each field test, we are interested in the questions of how these technologies can help to reduce cognitive load during wayfinding, and how these technologies influence the acquisition of spatial knowledge. This paper will report the methodology and results of the first experiment, with a focus on comparison of spatial knowledge acquisition with these interface technologies.

## 11.2 Study Design

### *11.2.1 Study Route and Participants*

A route in the city centre of Salzburg (Austria) was selected for the empirical test. It was divided into three sub-routes, each with nine decision points (e.g., intersections where multiple outgoing choices exist). The surroundings of these sub-routes are characterized by residential and business areas.

Twenty-four participants took part in the study (12 female and 12 male). The mean age was about 40 years (range 22–66). They were paid for their participation. All of them are German-speaking people.

### *11.2.2 Navigation Prototypes*

For studying wayfinding performance and spatial knowledge acquisition with different interface technologies, we used three self-implemented mobile navigation prototypes running on Apple's iPhone 4. These prototypes used map-based, AR-based, and voice-based interfaces respectively. Recent findings on pedestrian navigation from literature were integrated and considered when developing these prototypes.

Figure 11.1 shows a screenshot of the map-based prototype. The route is visualized as a red line filled with small white arrows pointing the forward direction. The past path is dyed in a lighter colour to be clearly separated from the future path. The real-time position is determined by GPS, improved by a route matching algorithm. A "track-up" egocentric map view is provided and its centre is adapted automatically to the current location. When a user is close to a decision point, a semi-transparent, blue-white directional arrow appears at the bottom-right of the screen. The arrow shows the directions based on the 7-sector model proposed by Klippel et al. (2004). Some other functions are also provided, such as zooming and panning.

In the AR-based prototype, route information is overlaid on the real world camera view. GPS module, magnetometer, and the tilt sensor on the mobile devices are used to calculate the position of the overlay information. Depending on the distance to the next decision point (DP), the overlay information changes from a red circle marking the position of the DP, to a bigger red circle showing remaining distance to the DP, and finally to a bigger red ring enclosing the waypoint to be entered. By changing the style and enlarging the size of the overlay, we expect users to get a feeling of crossing a portal. In additional to the graphical interface, a vibration alarm is raised when a decision point is reached. A screenshot of the AR-based prototype is given in Fig. 11.2.

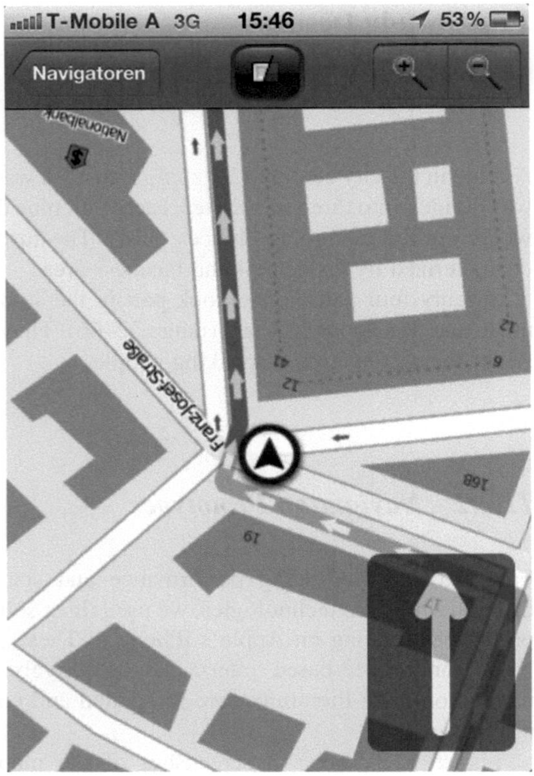


Fig. 11.1 A screenshot of the map-based prototype, with an egocentric view, distinction between past and future path, turn direction arrow, and zooming and panning function, etc.

**Fig. 11.2** A screenshot of the AR-based prototype, showing a ring. The distance to the next decision point is shown on the left of the ring to indicate a left turn at the current decision point

The development of the voice-based prototype was based on the semantic-based model developed in the previous project SemWay (Rehrl et al. 2010). Based on the model, semantic-based route instructions instead of metric-based route instructions can be provided, for example, “walk straight, pass the theatre, and walk to the crossing” instead of “walk straight for 103 m”. Verbal instructions for each decision point of the test route were automatically generated by using the semantic-based model (Rehrl et al. 2010). The interface of the voice-based prototype includes a single screen with a slider for controlling the sound volume and a button for repeating the last instruction. When a user gets close to a decision point, the mobile device vibrates, and plays the voice instruction describing the actions from this decision point to the next.

### *11.2.3 Design and Procedure*

Participants were randomly divided into three groups. A within subject design and a counterbalancing consideration were used for the test, i.e., for each sub-route, these three groups each used one of the navigation prototypes (mobile map-based, AR-based, and voice-based). When they reached the next sub-route, they switched to another prototype. Each participant was accompanied by two researchers. One observed the test run and guided through the interviews and the other collected quantitative and qualitative performance measures (e.g., the number of stops, duration of stops, and reasons of stops). Participants’ movement, interaction with the navigation prototypes, task completion time, and GPS accuracy were also logged on mobile phones.

At the beginning of the test session, we explained the basic usage of the pedestrian navigation prototypes to the participants and gave a short demonstration of the prototypes on the mobile phones. After a brief training session, participants were led to the starting point of the first sub-route. The task for them was to navigate to the end of the sub-route. If participants decided wrongly at a decision point, the observing researcher used gestures to indicate the correct choice. No other assistance was given during navigation. In order to avoid any influence on participants, the researchers walked several metres behind participants. When reaching the end of the sub-route, participants were asked to answer questionnaires assessing usability and task load, and to give some further qualitative feedback and experiences. In addition, they were asked to solve some tasks assessing spatial knowledge acquisition:

- Pointing task: to give an approximate direction to the starting point of the current sub-route, measured in degrees via a digital compass on mobile phones;
- Sketching map: to draw a sketch map of the area they just passed as precisely as possible, focusing on the route and landmarks;
- Marking task: to mark the half of the sub-route on their sketch maps;
- To indicate their familiarity with the current sub-route before the test.

When finishing all these tasks, participants switched to another prototype, and the same procedure was repeated for the next sub-route. In total, the test for each participant was completed within 1 h.

## 11.3 Results and Discussion

The field experiment was completed on Nov. 2010. All the participants successfully finished the tasks. The results of the experiment include two parts: wayfinding performance and user experience, and spatial knowledge acquisition. Results assessing wayfinding performance and user experience can be found in Rehrl et al. (2011). In this paper, we report the results of spatial knowledge acquisition.

The results on the aspect of spatial knowledge acquisition were analysed by focusing on sense of direction (the pointing task), sketch maps (topological aspects: sketched landmarks; errors in sketching turns, i.e., missing/wrong/unnecessary turns), and sense of distance (marking half of the route). We only considered the results from participants who were unfamiliar with the sub-routes. In total, we got 24 participant/sub-route pairs (8 for mobile maps, 8 for AR, and 8 for voice).[1] The male–female ratios were similar in the three groups. In the following, we present and discuss the results.

### *11.3.1 Sense of Direction (The Pointing Task)*

The sense of direction was measured as the deviation between actual directions and pointed directions in the pointing task. The deviations were measured in degrees. Figure 11.3 shows the results of sense of direction, comparing mobile maps, AR, and voice.

The results show that map users and voice users performed considerably better in pointing to the start compared to AR users. The results for AR users were not surprising, because the AR-based prototype suffered from the poor GPS signal (in both the map-based and voice-based prototypes, some route matching algorithms can be used to improve the GPS accuracy) and poor compass accuracy, and thus brought some confusion to the users. However, map users did not perform considerably better than voice users, which is inconsistent with our expectation. A possible explanation is that map users did not make full use of the map-based prototype, e.g., according to the usage log, they seldom used the zooming function to get an "overview map". We also did a one-way ANOVA (analysis of variance)

[1] The spatial knowledge acquisition test was conducted within a framework including many other tests, in which familiarity was not the only criterion in choosing participants. Therefore, only 24 participant/sub-route pairs were "unfamiliar".

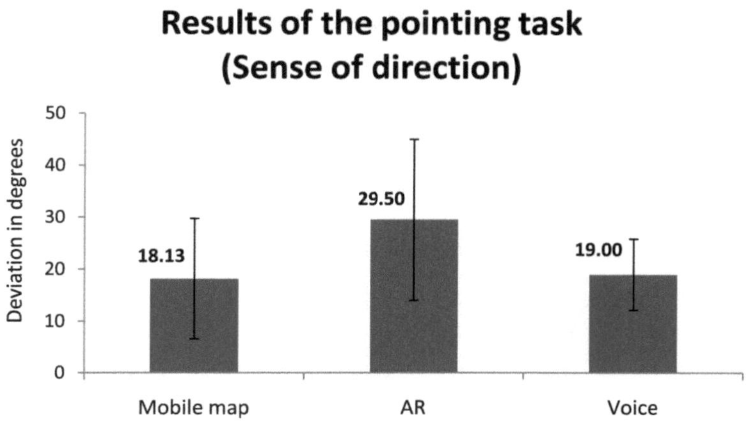


**Fig. 11.3** How the sense of direction differs among different navigation conditions. *Vertical bars* denote 95% confidence intervals

test.[2] According to the test, the difference among these three navigation conditions was not significant (F(2,21) = 1.60, p = 0.23).

## *11.3.2 Sketched Landmarks*

Literature has shown that it is useful to stick to topological interpretation of sketch maps only (Lynch 1960; Gartner and Hiller 2009). Therefore, the analysis of sketch maps focused on two topological aspects: sketched landmarks (landmark names), and errors in sketching turns. The later aspect will be discussed in next section. In this section, the results of sketched landmarks are presented. Figure 11.4 shows the mean number of sketched landmarks in these navigation conditions.

According to Fig. 11.4, voice users drew more landmarks in their sketch maps compared to AR users and map users. We also found that 78% of the landmarks sketched by voice users were mentioned/included in the verbal wayfinding instructions. Therefore, the reason why map users and AR users drew fewer landmarks may be that: for the map-based and AR-based prototypes, landmarks were not explicitly highlighted (e.g., they were displayed in the background map in the map-based prototype, and were not visualized in the AR-based prototype). While for the voice-based prototype, landmarks were explicitly included in the verbal instructions. However, there was no significant difference in the number of sketched landmarks across different navigation conditions (F(2,21) =1.63, p = 0.22).

[2] The surroundings of each sub-route are characterized by residential and business areas. In addition, the three sub-routes have comparable complexity (in terms of the number of decision points). Therefore, we did not consider sub-routes as a factor when comparing the performance among different navigation conditions, and a one-way ANOVA was used.

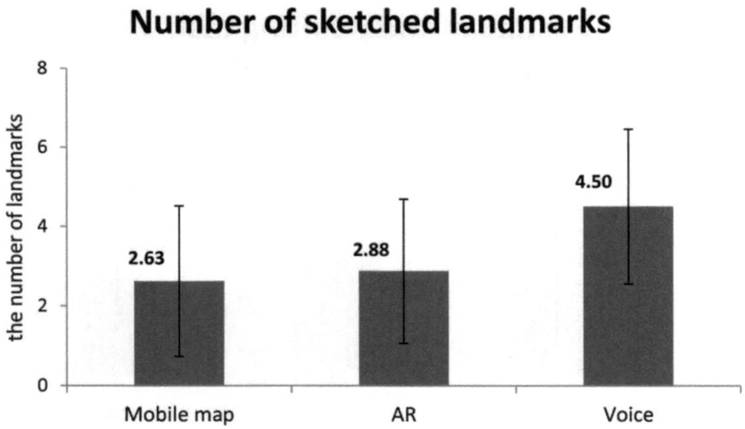


**Fig. 11.4** How the number of sketched landmarks differs among different navigation conditions. *Vertical bars* denote 95% confidence intervals

### *11.3.3 Errors in Sketching Turns (Missing/Wrong/Unnecessary Turns in Sketch Maps)*

The other aspect from the analysis of sketch maps was errors in sketching turns. We compared each sketch map with the actual route map to check whether there were any missing turns, any wrong turns (e.g., right turn in the actual route map, while left turn in sketch maps), and any unnecessary turns. Each missing/wrong/unnecessary turn was counted as an error in sketching turn. Figure 11.5 shows the mean number of errors in sketching turns.

The results shows that map users made considerably fewer errors in sketching turns compared to AR users and voice users. This is consistent with our expectation: in the AR-based and voice-based prototypes, turns were not conveyed/presented in a spatial-related overview context. As a result, AR users and voice users would make more errors in sketching turns. However, the difference among these three navigation conditions was not significant ($F(2,21) = 1.23$, $p = 0.31$).

### *11.3.4 Sense of Distance (Marking the Half of the Sub-routes)*

In the test session, we asked participants to mark the half of the sub-routes on their sketch maps. A grading system was developed to measure these marks. Every sub-route was equally divided into 20 segments, which were named as “1”, “2” … “9”, “10”, “10”, “9” … “2”, “1” respectively. A participant’s mark was graded as the name of the segment where it was located. Therefore, the grades scaled from 1 (worst) to 10 (best). Figure 11.6 shows the mean grade for each navigation condition.

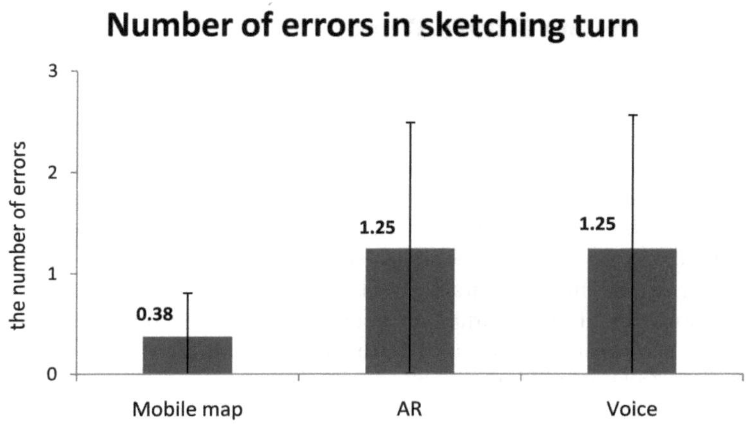


**Fig. 11.5** How the number of errors in sketching turns differs among different navigation conditions. *Vertical bars* denote 95% confidence intervals

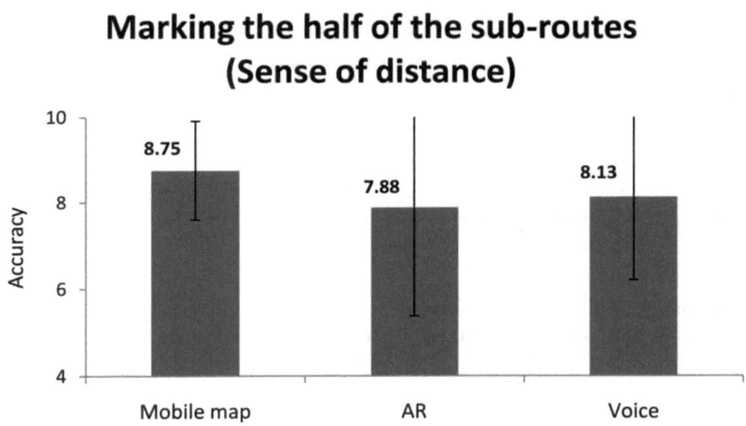


**Fig. 11.6** How the sense of distance differs among different navigation conditions. *Vertical bars* denote 95% confidence intervals

The results in Fig. 11.6 show that map users performed better in marking the half of the sub-route compared to AR users and voice users. However, the differences among them were not significant as we expected. An explanation may be that map users did not make full use of the map-based prototype. As a result, for all three navigation conditions, the knowledge about sense of distance was mainly gained from sensual perception of the real world , but not from the navigation prototypes. The comparable results of three navigation conditions might be also due to the poor differentiation of the grading system.

According to the ANOVA test, the difference among these three navigation conditions was not significant ($F(2,21) = 0.30$, $p = 0.74$).

## 11.4 Summary and Future Work

This paper reported a work in progress on comparing spatial knowledge acquisition with mobile maps, augmented reality (AR), and voice in the context of GPS-based pedestrian navigation. Results of a field test were presented and discussed. In summary, the current results show that, among different navigation prototypes (map-based, AR-based, and voice-based), using the map-based prototype leads to more accuracy in pointing to the start (sense of direction), more accuracy in sketching turns, and more accuracy in marking the half of the route (sense of distance). However, it is important to note that no significant differences were found for the above aspects. This may be due to the small set of test persons. More research needs to be done on this issue.

Currently, we are planning another field experiment to investigate the differences of spatial knowledge acquisition with mobile maps, AR, and voice in more details. More participants who are unfamiliar with the environment will be recruited. In addition, we will differentiate three kinds of spatial knowledge, namely landmark, route, and survey knowledge (Siegel and White 1975), and study how these interface technologies influence the acquisition of each of them.

The field test also opens several important research questions for future navigation system development: "do users care about spatial knowledge acquisition during navigation", and "if yes, how can we design a navigation system that not only reduces users' cognitive load during wayfinding, but also enables them to acquire spatial knowledge". Related findings of spatial cognition and human wayfinding together with usability studies need to be integrated to address these questions.

**Acknowledgements** This work was supported by the ways2navigate project, funded by the Austria FFG's ways2go programme (2009). We thank our partners, namely Salzburg Research, FACTUM, TraffiCon and WalkSpace Mobilität, for their contributions to the project.

## References

Aslan I, Schwalm M, Baus J, Kruger A, Schwartz T (2006) Acquisition of spatial knowledge in location aware mobile pedestrian navigation systems. In: MobileHCI'06. Espoo, ACM Press, New York, pp 105–108

Gartner G, Hiller W (2009) Impact of restricted display size on spatial knowledge acquisition in the context of pedestrian navigation. In: Location Based Services and TeleCartography II, Springer, pp 155–166

Ishikawa T, Fujiwara H, Imai O, Okabe A (2008) Wayfinding with a GPS-based mobile navigation system: a comparison with maps and direct experience. J Environ Psychol 28(1):74–82

Klippel A, Dewey K, Knauff M, Richter K, Montello D, Freksa C, Loeliger E (2004) Direction concepts in wayfinding assistance. In: Baus J, Kray C, Porzel R (eds) Workshop on Artificial Intelligence in Mobile Systems 2004 (AIMS'04), SFB 378 Memo 84, Saarbrücken, pp 1–8

Krüger A, Aslan I, Zimmer H (2004) The effects of mobile pedestrian navigation systems on the concurrent acquisition of route and survey knowledge. In: Proceedings of Mobile HCI, Glasgow, pp 446–450

Lynch K (1960) The image of the city. MIT Press, Cambridge

Ortag F (2005) Sprachausgabe vs. Kartendarstellung in der Fußgängernavigation. Master thesis, Vienna University of Technology

Radoczky V (2004) Literature review and analysis about various multimedia presentation forms. Internal report in Vienna University of Technology

Rehrl K, Häusler E, Leitinger S (2010) Comparing the effectiveness of GPS-enhanced voice guidance for pedestrians with metric- and landmark-based instruction sets. In: Fabrikant S, Reichenbacher T, van Kreveld M, Schlieder C (eds) GIScience 2010. Springer, Berlin, pp 189–203

Rehrl K, Häusler E, Steinmann R, Leitinger S, Bell D, Weber M (2011) Pedestrian navigation with augmented reality, voice, and digital map: results from a field study assessing performance and user experience. In: Proceedings of LBS 2011, Springer

Siegel A, White S (1975) The development of spatial representations of large scale environments. In: Reese H (ed) Advances in child development and behaviour 10. Academic, New York, pp 9–55

Walther-Franks B (2007) Augmented reality on handhelds for pedestrian navigation. Master thesis, University of Bremen

Wikipedia (2011) Wizard of Oz experiment, http://en.wikipedia.org/wiki/Wizard_of_Oz_experiment. Accessed Aug 2011

# Chapter 12
# Developing Map Databases: Problems and Solutions

**István Elek and Gábor Gercsák**

**Abstract** This article is an introduction to a several-year long development focusing on the building of a digital map database, named EDIT (Elek I (2010) Nagyméretű térképi adatbázisok fejlesztési és működési tapasztalatai. In: Dövényi Z, Fodor I, Lovász GY, Schweitzer F, Tóth J (eds) Geográfia-2010-Pécs. Pécsi Tudományegyetem, Pécs). The theoretical background, the applied technology, the developed information system, and its perspectives will be described. The classical relation database technology and the object oriented, document based database management (Elek I, Giachetta R, Máriás ZS (2009) Egyetemi digitális térképtár fejlesztése az ELTE-n. In: Kákonyi Gábor (ed) Fény-Tér-Kép konferencia. Dobogókő, Hungary, http://www.geoiq.hu/index.php?option=com_docman&task=cat_view&Itemid=63&gid=35&orderby=dmdate_published&limitstart=15; Giachetta R, Elek I (2010) Developing an advanced document based map server. In: 8th international conference on applied informatics, Eger, Hungary, 27–30 Jan 2010; Marias ZS, Dezso B, Giachetta R, Elek I (2010) Cartographic symbol detection using local segmentation and other methods. In: Prasad B (ed) International conference on artificial intelligence and pattern recognition (AIPR-10), Paper 196, Orlando, 12–14 July 2010) are applied beside those web based tools and technological solutions which are in the world of the WEB2. The system is not only a rich data source for researchers but also a virtual laboratory of the database technology where various database management systems related to GIS can also be investigated. The EDIT system has become an ordinary data source of the research and education at Eötvös Loránd University and its partners in Hungary (Elek I, Kovács B, Verebiné Fehér K (2004) Digitális térképtár az ELTE-n, Magyar Földtudományi Szakemberek VII. világtalálkozója, Hungarian Geological Society, Budapest: 80). The role of the system is of remarkable importance in the doctoral training

---

I. Elek (✉) • G. Gercsák
Department of Cartography and Geoinformatics, Eötvös Loránd University, Budapest, Hungary
e-mail: elek@map.elte.hu; gercsak@ludens.elte.hu

L. Zentai and J.R. Nunez (eds.), *Maps for the Future*,
Lecture Notes in Geoinformation and Cartography 5,
DOI 10.1007/978-3-642-19522-8_12, 

too (Rigaux P, Scholl M, Voisard A (2002) Spatial databases with application to GIS. Morgan Kaufmann, Burlington).

## 12.1 Introduction

Research projects and teaching at the university missed the availability of digital maps for long, although there were many paper maps in the map collection. A large number of map databases are now accessible in the Internet: some of them are free, but they mostly charge for their use (http://lazarus.elte.hu/gb/linkek.htm, http://lazarus.elte.hu/hun/index.html, http://www.davidrumsey.com, http://www.lib.berkeley.edu/EART/topo.html, http://turistautak.hu, http://library.stanford.edu/depts/branner/collections/sovietmil.html, http://www.nasa.gov, http://www.terraserver.com). The laws regulating the use of maps are very different from country to country, and the data policy changes from government to government. It is not surprising that "sharp" maps are rarely found on free web pages. As for the fresh satellite images and aerial photos, the case is the same. It is even more difficult to find free vector maps, particularly those that have link codes to databases.

University researchers and teachers very much need a well-organized cartographic database that is available for everyone at the university. The Department has done important steps in this direction by setting up an information system named EDIT. (The acronym stands for the Hungarian words of University Digital Map Collection.) More than 10,000 maps have been entered into the database, which is completed with a data loading application and an online query system (Fig. 12.1). Although the original plan only counted with users from the university, and the data loading is not complete yet, and a part of the users of the system already comes from outside the university. In the beginning, the EDIT system was only accessible from the nodes in the *elte* domain that is for the teaching and research units that take part in the training of students of earth sciences and computer science. Later, the range of users greatly expanded. The University of Debrecen and the University of Pécs were the first to join the system. Further members include the University of Szeged, two Academic institutions (the Geographical Research Institute and the Balaton Limnological Research Institute), and the Institute and Museum of Military History.

This paper introduces the EDIT system and its major parameters.

## 12.2 University Digital Map Collection (EDIT)

The EDIT is an information system built on the principle of relation databases and made for the management, inventory, and service of raster and vector map data. The raster data include scanned maps, orthophotos, and hyperspectral aerial images, while the vector data are mostly available in ESRI shape and Mapinfo tab formats.

**Fig. 12.1** First page of the online query system

The logic and the background structure of the system, and the applications of the database provide the users with an easy and fine-tuned query, and a quick presentation of the results (alphanumerical and graphical or map data). Further, the access to the system can be easily regulated.

The system consists of two major parts: the EDIT is for data loading and maintenance, while the EDITKE is for online query (Fig. 12.1). The online query system is accessible at http://mapw.elte.hu/edit for those on the *elte* domain. The clients of the Virtual Private Network (VPN) can reach this URL through the certificate of the VPN. The data maintenance application runs on the map server of the system. The language of the system is Hungarian.

The system was made in Microsoft VisualStudio 2008. The loading application is a win32 application. The online query was made in ASP.NET. István Elek developed both application programs.

## 12.3 Structure of the Database

The great differences between the formats of the raster and vector maps caused difficulties for the planning of the database. Although there are various raster formats, they all contain an *m* x *n* matrix, which stores the intensity values of a picture according to a colour model in rows and columns. These data may come from scanned paper maps or from the signals of digital images. They all follow the simple structure of raster data models. Descriptive data such as scale, time of making, sheet number can be added to the maps (namely, to each raster file).

This logic cannot be used for vector maps, because these maps are organized into layers and feature classes. Several types of information, layers, and feature classes together can only express the complexity of classical paper-based maps. In addition, some of the usual descriptive data (e.g., scale, title, and sheet number) cannot be interpreted for these data. It is obvious that different data structures must be used for storing the raster and vector maps due to their basic differences. Further, the vector data are often available organized in relation databases even though they are exported into file groups (e.g., ESRI shape or Mapinfo tab files). Consequently, the raster maps need a database structure completely different from that of the vector maps.

### *12.3.1 Storage of Raster Maps*

In most cases, the raster maps can be stored in a table structure in which one record represents one map with all its descriptive data as shown in the table below.

| map_id | Title | Sheet number | Scale | ... |
|---|---|---|---|---|
| 3,455 | Ödenburg | P48 | 300,000 | ... |
| ... | ... | ... | ... | ... |

The present large capacity database managers allow the binary storing of raster files, graphical data and various kinds of data in BLOB type fields. BLOB is an abbreviation of binary long object, which term expresses that it contains the original information converted into a series of bytes. To display the content of these fields is not as simple as that of the usual alphanumerical fields. However, a great advantage is its secure storage, because the authentication system of the database manager automatically protects the BLOB type data too. This is very important for the information systems that have a high risk of security.

However, the map collections do not belong to the category of high risk. In addition, it may also happen that the raster files of the maps must be modified (e.g., due to noise filtering, improving the quality of the image, georeferencing), when the storage in the BLOB fields would cause problems. Therefore, the maps are stored in another way. The descriptive data are stored in data tables, while the maps are stored in file systems outside the relation database. The descriptive tables contain the name of the maps only, which is a reference in the *map_id* field. Therefore, if the raster file is modified in any way, it will be immediately visible in the system without changing the database.

The raster map data are stored in two data tables. One stores the groups of maps (table of *groups*, Fig. 12.2), the other one stores the data records of maps (*rastermaps*, Fig. 12.3). Individual maps are sorted into groups by referring to the table of *groups*.

This simple relation (of *1:n type*) link describes the grouping of maps. Several other field values may be chosen for this purpose from the collection (e.g., scale), but as the present project is a pilot plan, we decided to have a simple structure first.

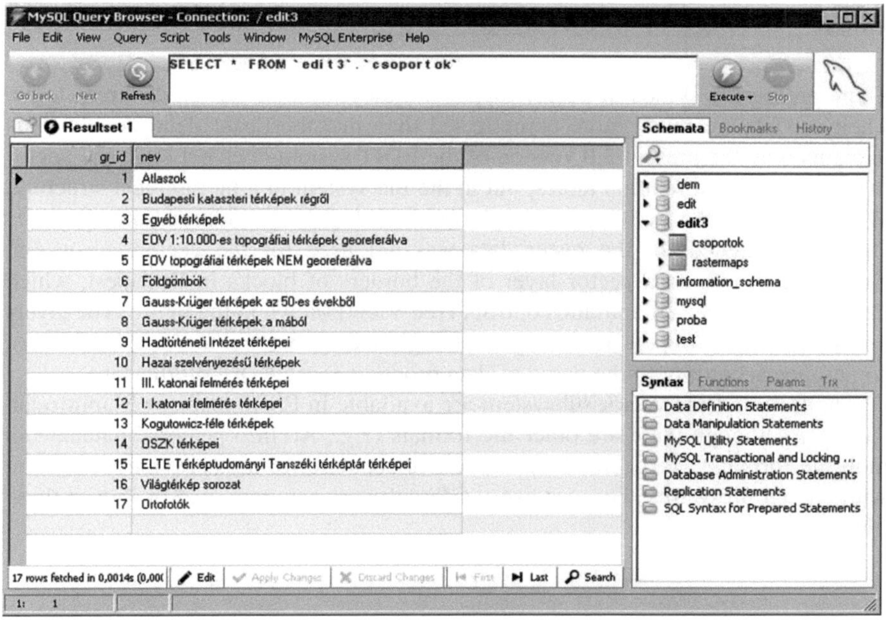


**Fig. 12.2** Table of *groups*

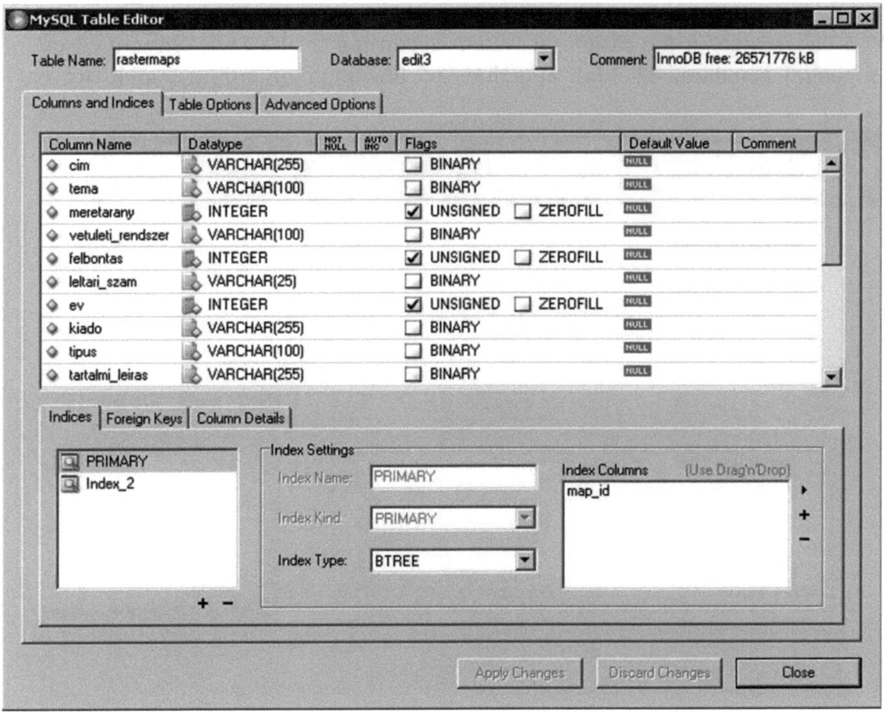


**Fig. 12.3** The structure of table *rastermaps*

### 12.3.2 Storage of Vector Maps

There are various kinds of vector maps, and they are built up on different logic. Their models are much more complicated than that the raster data models. This explains why the present 3.0 version of the EDIT system does not store the vector maps organized in relation tables, but in the file system in a hierarchical structure. When planning the system, its simple use was highly considered, and the table structure designed for the raster data was not used for the vector subsystem. Figure 12.4 presents a vector layer of the borders of blocks in Budapest, which can be accessed by a hierarchic control (Tree View) shown in the figure. The layers are sometimes grouped according to a region they belong to (e.g., *budapest*), sometimes according to the name of data products (e.g., ADC-WorldMap).

The layers of the vector subsystem are available in ESRI shape or Mapinfo tab formats. Although there are other file formats (e.g., ArcInfo personal database in mdb file formats), we decided that these two standard formats provide the system with enough flexibility, because every GIS software can now read or import these file formats.

### 12.3.3 Query from the Database

You can search in the raster subsystem according to the SQL (structured query language) standard. The full functions of this kind of online search are only

**Fig. 12.4** Navigation in the vector subsystem with a hierarchic control

available for special users with the help of a series of SQL commands and to a very much limited extent for users without personal identification (such as queries from any *elte* domain). Naturally, the whole arsenal of SQL can be applied in the process of data loading and database building, because the data loading application supports these processes.

There is no SQL interface in the vector subsystem. The expressive names of libraries and files make the navigation easy in the file system. A table of metadata might be put into service with the pure task of querying in the future, but it is certainly not going to be a part of version 3.0.

## 12.4 Authentication, Security and Data Loading

The authentication of the database against hackers in the cyber space is a very important issue. It is not worth building large databases if we cannot protect the server from the attacks of hackers, because the frequency and complexity of the attacks will surely lead to data loss.

The protection must be guaranteed at two levels. Namely, the server and database must be protected. The protection of the server is the task of the administrator of the operating system, which includes the following tasks: virus protection, setting up of the firewall, setting up of the Internet Information Service, and the regular refreshing of the operating system of the server. The server is not accessible from a local network. It has only one public directory, into which certain files (e.g., tif, jpg, tab, shp) can be loaded from a few dedicated computers. In this case, the data loading application imports the requested maps.

Although the protection of the server protects the database too, but the efficient protection of the database needs additional settings. It must be decided in the administration of the database management system that which users have the right to access the database tables and to what extent. There are users who, without personal identification, have the select right only. (These are requests from URLs on the *elte* domain.) Others, who have personal authentication, have not only the *select* right to access the database. The series of SQL commands is also available for them, and they can download the selected maps on their client computer.

Only users with special rights are allowed to load and maintain the data. They have the right to not only select, but also to update and delete information. However, they do not have the right to delete large amount of data, to drop tables or to perform any other operation that would corrupt the data. The administrator only has full rights over the database.

The loading and maintenance of data is performed by a win32 application running on a server, which is available for registered users only. This application allows the editing of the existing map data, the changing of the attribute data, as well as updating the data and the manipulation of the existing maps (Fig. 12.5). In addition, the new maps are entered into the system by using this program. The directory of the server that receives the uploaded data is only accessible with the

**Fig. 12.5** Importing groups of maps into the system

right of writing from a few client computers. The map data to be uploaded into the database are placed here, in the buffer. After loading the map into the database, the buffer becomes free. The configuration file of the system includes, among others, this place as well, which is specified at the setting of the system.

## 12.5 Query and Servicing

The system includes a configuration file, which contains the parameters necessary for the operation. This file ensures that the places that are needed for the access to and management of the data and maps are flexibly specified. This guarantees the smooth running of the system even if the operation system is completely reconfigured. In this case, the accessibility and addresses of the modified places have to be rewritten in the config file only. There is another parameter file, which gives the names of the data columns to be displayed online. This is necessary because not every user is interested in every data column. This parameter file allows us to specify which data columns should be visible for the online query.

When uploading data, complicated queries may be necessary. This is offered by a series of SQL commands, which allows the user to make a combination of SQL commands at wish (Fig. 12.6).

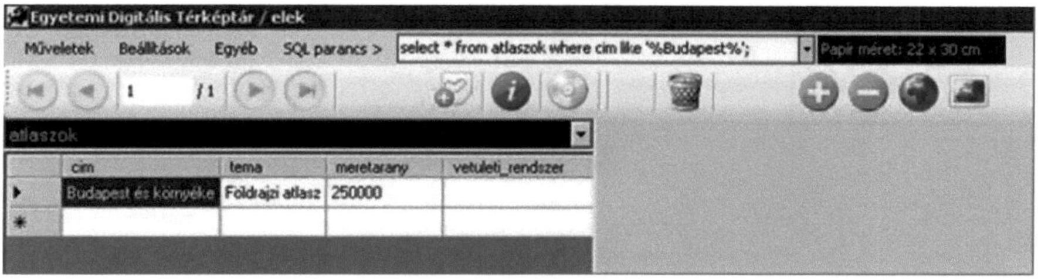


**Fig. 12.6** Searching for maps that have the word *Budapest* in the title

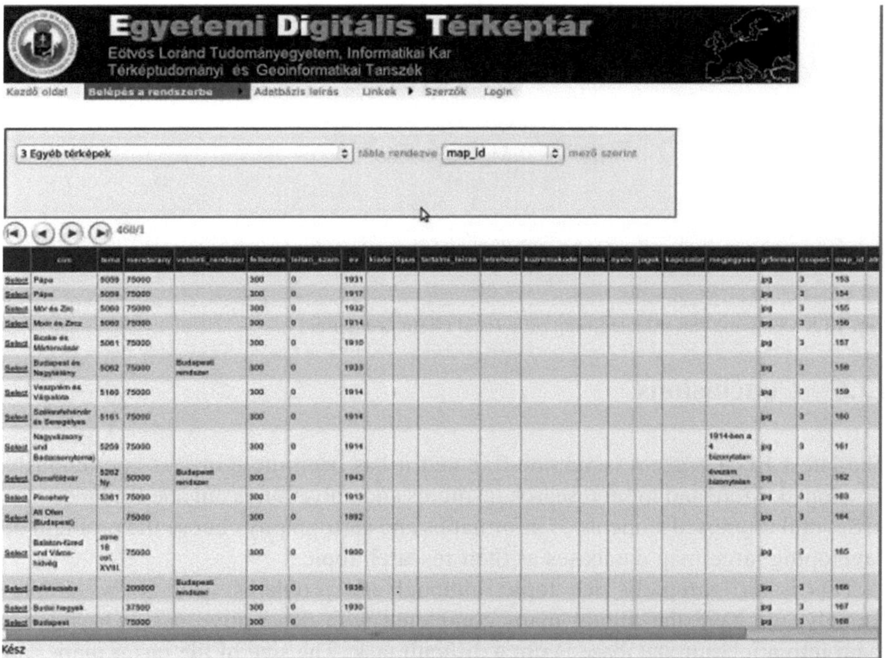


**Fig. 12.7** The attribute data of the first 20 maps in the *egyéb (other)* group of maps arranged according to the field *map_id*

The data uploading application offers a function that allows us to copy the selected map into a specific place. (Its role is very much similar to that of the *Download* function of the online query.) The user can copy the map into any place from here. One of the major aims of EDIT system is to provide the researchers and students with free access to any map stored in the system. This place, among others, is included in the configuration file of the system.

The web and the database technology together led to the building of an online search and query system, EDITKE (KE stands for the Hungarian word *keres*). The program in ASP.NET is an application of the server, which displays the data of the maps that satisfy the given conditions (Fig. 12.7) and the selected map (Fig. 12.8).

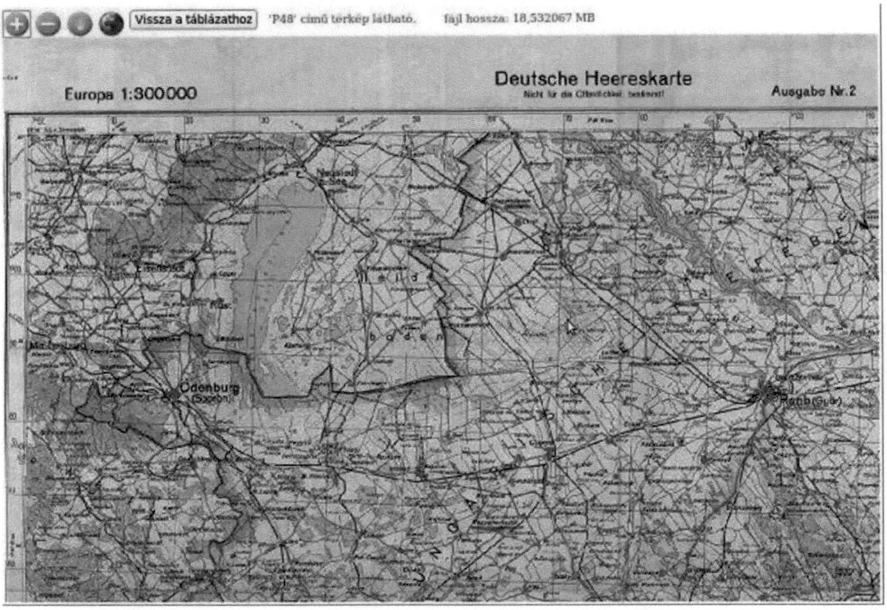


**Fig. 12.8** Viewing the selected map on a local client computer

## 12.6 Conclusions

The latest EDIT system is version 3.0, and it has been more or less continuously developed. More and more research institutes and universities outside Budapest use it. Several interested students of informatics and cartography choose the problem of developing large map databases as their research topic.

This is an extremely rich topic, although the problems seem to be simple. Namely, we have the digital maps completed with descriptive data. Organizing them into a relation database is not a difficult task. The size of the raster maps can be very large (between 25 and 500 MBytes), while the vector maps have very varied structures. The size of a hyperspectral image can be extremely large (even 4–4.5 GBytes). As a result, the total size of the images for a certain area may exceed 1 TByte (1,000 GByte). Handling such a large amount of data is not easy at all, not to mention displaying them. Displaying the images of several megabyte size on the Internet is particularly problematic, and moving these images on the web is really difficult. It is necessary to develop special programs (services) on the server side that generate such images the size of which do not exceed the size offered by the resolution capacity of the given video card. This is the way to eliminate the barriers caused by the differences in the bandwidth.

The development must consider some other aspects that are not included in the present 3.0 version. One of these aspects is the transfer to or at least the study of transferring to object-oriented database management (e.g., the use of MongoDB) in the map server. Another aspect is the tracking of changes in time. The relation database managers always face the difficulty of tracking the changes in time, particularly if complicated graphic data (such as maps) change.

The present EDIT system runs in Windows server environment (Windows server 2008, Internet Information Service, VisualStudio, ASP.NET), but it would be interesting to test it in Linux environment (Linux, Apache, Postgresql, Java). The increasing popularity of open source systems cannot be stopped: therefore, limiting the system to Windows environment would not follow the trend.

These are important questions, because the EDIT map collection is an experimental system, the terrain of informatical and cartographic experimenting, where all kinds of new technologies, algorithms and ideas can be tested.

**Acknowledgement** The European Union and the European Social Fund have provided financial support to the project under the grant agreement no. TÁMOP 4.2.1./B-09/KMR-2010-0003.

# Chapter 13
# The Tile-Based Mapping Transition in Cartography

**Michael P. Peterson**

**Abstract** Arguably, the major development of the first two decades of maps and the Internet is a method of map distribution that divides the map into smaller tiles. The tiling of maps along with a more interactive form of communication with a server called AJAX transformed the online mapping experience. Soon after the method was introduced through Google Maps, all other online map providers switched to the new form of map distribution. The introduction of Application Programmer Interfaces that allowed user information to be added to the maps solidified this form of online map presentation. Whether for good or bad, online mapping is currently in a tile-based era and will likely be so for the foreseeable future. The method is examined more closely along with the potential for adding user-defined maps.

## 13.1 Introduction

Maps became a major component of the online information landscape in 1993 after graphical illustrations could be added to World Wide Web pages (Peterson 2003). All types of maps became easily available, from subway maps of cities to maps of the moon. Map users could even choose between alternative representations of the same environment and determine the one that best suited their needs. Interactive sites made it possible to center the map on an area of interest and include features requested by the user.

The nearly two decades since 1993 can be seen as a period of discovery as map providers tried different ways to make use of the new medium. Paper maps were initially scanned. Then, cartographers alternated between various vector and raster formats and forms of interactivity. A major goal was to increase the speed of map

M.P. Peterson (✉)
Department of Geography/Geology, University of Nebraska at Omaha, Omaha, NE, USA
e-mail: mpeterson@unomaha.edu

L. Zentai and J.R. Nunez (eds.), *Maps for the Future*,
Lecture Notes in Geoinformation and Cartography 5,
DOI 10.1007/978-3-642-19522-8_13, 

delivery. Map users did not want to wait for the map and a method was devised to appease the impatient online map user. This method involved dividing the map into pieces, or tiles. It is important to understand the tile-based mapping transition and what it means to cartography. We begin by examining server-based mapping.

## 13.2 Server-Based Mapping

Interactive server-based mapping began in 1993 within months after the introduction of the Mosaic browser in March of that year. The first interactive online mapping application was developed by Steve Putz (1994) at the Xerox Palo Alto Research Center (PARC). His Map Viewer program allowed the user's client computer to create on-demand maps from a geographic database. Each interaction with Map Viewer would request a new map from the server that was zoomed in on a specific point (see Fig. 13.1). Individual maps were generated in a graphic file and embedded into a web page.

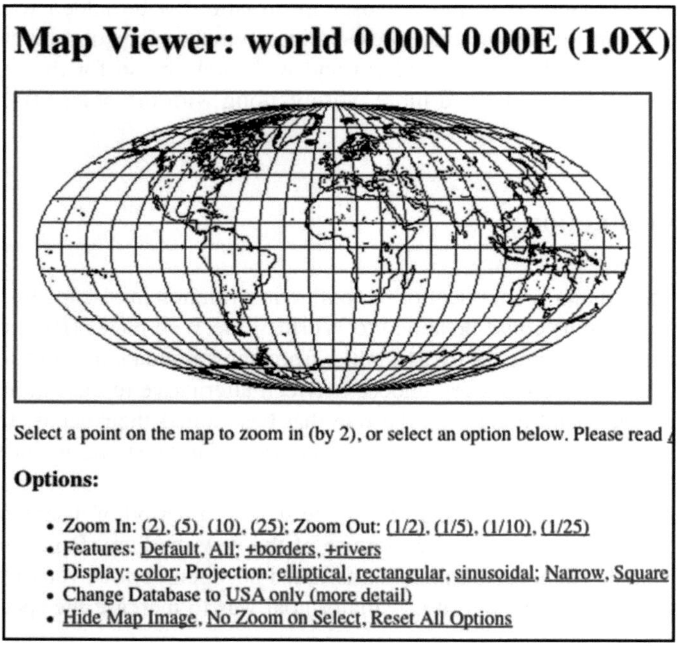


**Fig. 13.1** Xerox Parc Map Viewer was an early example of an interactive web map. By interacting with a mapping program on the server, the site made it possible to generate a map of the world at different zoom levels. The resultant map was converted into a graphic file and inserted into a web page

MapQuest introduced its online mapping site in 1996 based on a large database that included most streets in the North America. Within months, it was producing millions of maps every day and was the leading online map provider until 2009. The user's client computer would make a request for a specific map. MapQuest servers would respond to the request by drawing the map from a database of points and lines, converting this to a grid-based raster format and delivering the resultant map within a web page.

Each request for another map, at a different zoom level or centered at another point, would result in another server request that would produce another map that would be embedded in another webpage that would update the page on the user's computer. Although the process was fairly fast, there was always a wait for the server to respond. A simple zoom or pan required waiting for the server to produce another map that was inserted into another webpage. Server requests are also subject to Internet traffic so a request for a map might take considerably longer when traffic was heavy. Maps would be produced more quickly in the overnight hours when Internet traffic was lighter. This variability in response times was found to be more annoying for users than the generally slow response times. Figure 13.2 depicts a 2001 version of MapQuest along with three maps of southern Florida at different scales.

## 13.3 Tiled Web Maps

Google Maps, introduced in 2005, offered a more interactive street map interface (Peterson 2007). Google, known for its search engine, effectively added a map-based search engine through Google Maps and the stand-alone Google Earth. In the process, they found a more effective way to indirectly make money from online maps by charging businesses, much like the way they make a profit through their search engine. In addition, by not including ads around the map, like MapQuest, they left more room for the map on the computer screen. Google Maps is based on two major ideas: (1) image tiling; and (2) AJAX.

### *13.3.1 Image Tiling*

Image tiling had been used since the early days of the World Wide Web to speed the delivery of graphics (Sample and Ioup 2010). In comparison to text, images require more storage and therefore take longer to download. A solution was to divide the image into smaller segments, or tiles, and send each tile individually through the Internet. These smaller files often travel faster because each can take a different route to the destination computer. On the receiving end, the tiles are reassembled in their proper location on the web page. With a moderately fast Internet connection, all of this occurs so quickly that the user rarely notices that

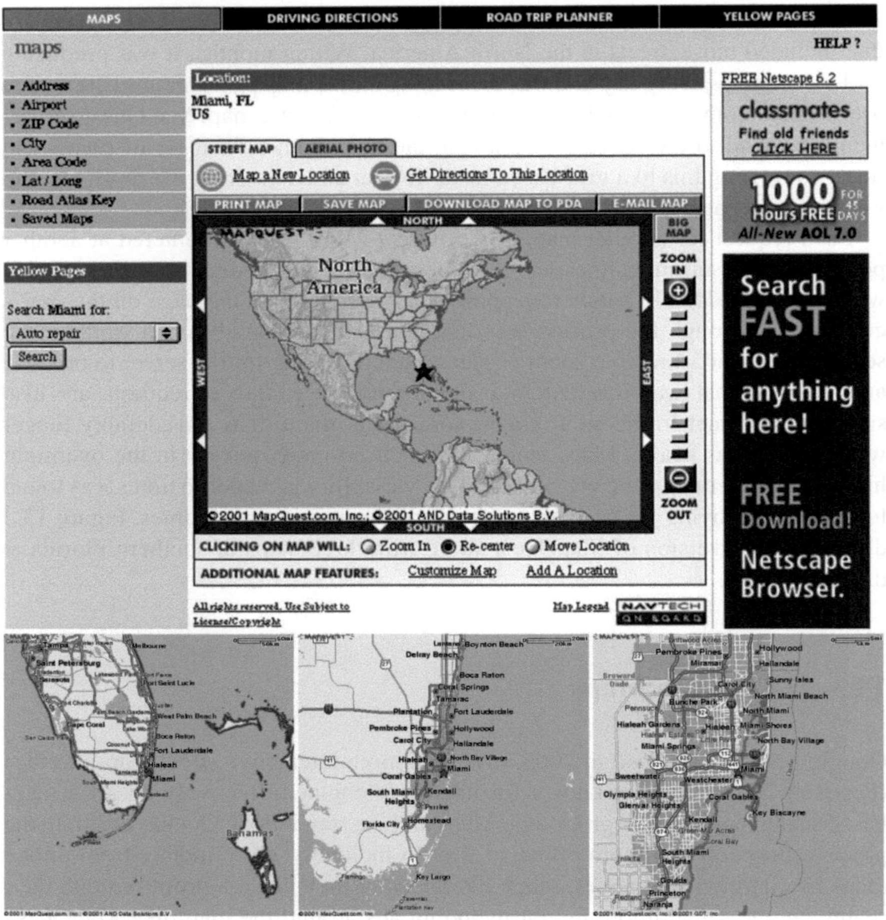


**Fig. 13.2** A 2001 version of the MapQuest webpage. Dominated by ads, the map constitutes only a small part of the webpage. Three different maps of southern Florida are also depicted. A total of ten zoom levels were available

the image is actually composed of square pieces. With a slower Internet connection, the individual tiles are clearly evident.

The basic street map is the most functional of all the views provided by these services. The map is provided in up to 20 scales called levels of detail (LOD). Each map at a particular level of detail is rendered from an underlying vector database consisting of points, lines and areas. This map is converted into a grid representation by placing the points, lines, and areas into a matrix of pixels. In the process, anti-aliasing is performed by adding lighter pixels around sharp edges to soften the stair-step effect. After the map is placed in a matrix that can be millions of pixels on each side, it is divided into tiles—usually 256 × 256 pixels (see Fig. 13.3). This process is repeated for every LOD.

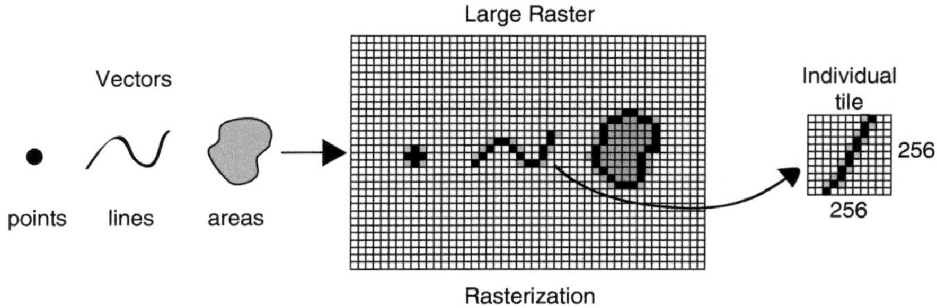


**Fig. 13.3** The process of making map tiles. A database of points, lines and areas is rasterized into a large matrix that is subsequently divided into 256 × 256 square tiles

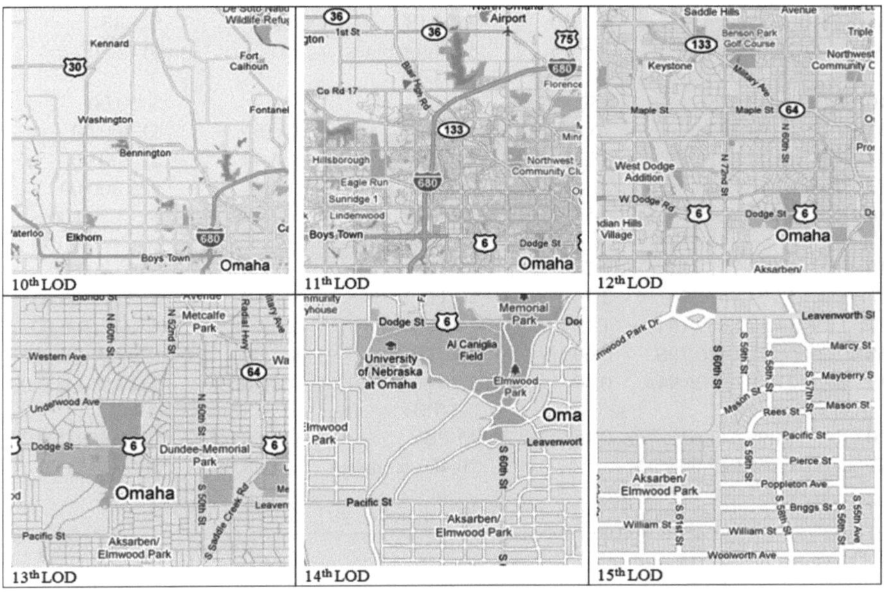


**Fig. 13.4** Individual map tiles from Google Map at six different levels of detail (zoom levels). In 2005, Google introduced a tiling system to deliver online maps. Over a trillion tiles are used for Google's 20 zoom levels

Figure 13.4 depicts a series of Google Map tiles at different scales. All tiles are 256×256 pixels and require an average of 15 KB a piece to store in the PNG format. Table 13.1 shows the number of tiles that are used in a tile-based mapping system for 20 levels of detail (LOD), or zoom levels, and the associated storage requirements and storage costs. With 20 LODs, there are a total of approximately 1 trillion tiles for the whole world. At an average of 15 KB per tile, the total memory requirements would be 20 Petabytes, or 20,480 Terabytes. No single client

**Table 13.1** The number of tiles, storage requirements, and storage costs used by a tile-based online mapping system to represent the world at different levels of detail (LOD) or zoom levels

| Levels of detail (LOD) | Number of tiles | Ground distance per pixel in meters | Storage requirements at 15 kilobytes per tile | Disk storage costs at $100 per terabyte | RAM memory storage costs at $30 per gigabyte |
|---|---|---|---|---|---|
| 1 | 4 | 78,272 | 60 kilobytes (KB) | $0.000006 | $0.002 |
| 2 | 16 | 39,136 | 240 KB | $0.00002 | $0.007 |
| 3 | 64 | 19,568 | 968 KB | $0.0001 | $0.03 |
| 4 | 256 | 9,784 | 3.75 megabytes (MB) | $0.0004 | $0.11 |
| 5 | 1,024 | 4,892 | 15 MB | $0.001 | $0.44 |
| 6 | 4,096 | 2,446 | 60 MB | $0.006 | $1.76 |
| 7 | 16,384 | 1,223 | 240 MB | $0.02 | $7.03 |
| 8 | 65,536 | 611.50 | 960 MB | $0.09 | $28.13 |
| 9 | 262,144 | 305.75 | 3.75 gigabytes (GB) | $0.37 | $112.50 |
| 10 | 1,048,576 | 152.88 | 15 GB | $1.46 | $450.00 |
| 11 | 4,194,304 | 76.44 | 60 GB | $5.86 | $1,800.00 |
| 12 | 16,777,216 | 38.22 | 240 GB | $23.44 | $7,200.00 |
| 13 | 67,108,864 | 19.11 | 968 GB | $93.75 | $28,800.00 |
| 14 | 268,435,456 | 9.55 | 3.75 terabytes (TB) | $375 | $115,200.00 |
| 15 | 1,073,741,824 | 4.78 | 15 TB | $1,500 | $460,800.00 |
| 16 | 4,294,967,296 | 2.39 | 60 TB | $6,000 | $1,843,200.00 |
| 17 | 17,179,869,184 | 1.19 | 240 TB | $24,000 | $7,372,800.00 |
| 18 | 68,719,476,736 | 0.60 | 960 TB | $96,000 | $29,491,200.00 |
| 19 | 274,877,906,944 | 0.30 | 3.75 petabytes (PB) | $384,000 | $117,964,800.00 |
| 20 | 1,099,511,627,776 | 0.15 | 15 PB | $1,536,000 | $471,859,200.00 |
| Total | 1,466,015,503,700 | | 20,480 terabytes or 20 petabytes | $2,048,000 | $629,145,600 |

computer could have this much storage capacity. A data center is required to store this amount of data.

The cost of storing this much data on hard drives can be calculated based on a cost of about $100 for a 1 Terabyte drive, a price that does not include the housing or computer connection. To store the entire one trillion tiles would cost about $2 million ($100 × 20,480 Terabytes). In order to achieve faster response times, there is some indication that Google uses faster random-access memory (RAM) to store the Google map tiles. If the entire map of the world were stored in RAM, it would cost the company more than $629 million. It is possible that only a subset of tiles are stored with the rest being created on-the-fly based on demand.

These data storage requirements and costs are only for the map. The satellite view, with tiles in the JPEG format, requires approximately the same amount of storage space. A terrain view is also provided with 15 levels of detail. Google maintains multiple data centers around the world and each would likely have a copy of the map and satellite image, and any other map that is provided. Combining all of these data storage costs provides some indication of the importance placed on maps by Google and other companies. Even governments would have difficulty justifying the initial and ongoing expense of maintaining an online map in this way.

In a transformation that shows that imitation is the sincerest form of flattery, all of the other major interactive map providers—MapQuest, Yahoo, Microsoft (Bing)—converted from the standard server-client to the AJAX, tile-based method of map delivery within a short time after the introduction of Google Maps. In addition, they all used the same exact method of tiling with the same divisions. All of the tiles from the major map providers are inter-changeable—at least by position.

### 13.3.2 AJAX

The second major innovation brought by Google Maps was the incorporation of Asynchronous JavaScript and XML (AJAX) in the relationship between the server and client. This was the culmination of many years of effort to re-shape interaction on the Internet. Essentially, AJAX maintains a continuous connection with the server—exchanging small messages in the background even when the user has not made a specific request. This allows for faster server responses when the user does make a request. AJAX might be thought of an application that works in the background of a browser to anticipate what the user might want and be ready to communicate with the server to respond to a request. Operations in Google Maps that are particularly assisted by AJAX include zooming and panning, a common form of interaction with maps.

AJAX is a technique that combines JavaScript and XML to create very interactive, server-client web applications. AJAX is not a programming language in itself, but a term that refers to the use of a group of different technologies together. The technique uses a combination of HTML, Cascading Style Sheets (CSS), Document Object Model (DOM), and the eXtensible Markup Language (XML). These are all freely available technologies. Asynchronous communication is used to exchange data with the server while the user is idle so that the entire web page does not have to be reloaded each time the user makes a change (see Fig. 13.5). The result is increased interactivity, speed, and an improved user interface.

AJAX eliminates the usual start-stop-start-stop type of interaction. When the map is scrolled, additional map tiles are automatically downloaded. The tiles are added almost instantly because a connection is maintained to the server so that additional tiles can be quickly loaded. As the user scrolls, more of the map or satellite image is downloaded from the server without the user making a specific request for additional tiles.

Asynchronous communication is made possible by the AJAX engine, JavaScript code that resides between the user and the server. Instead of loading the webpage at

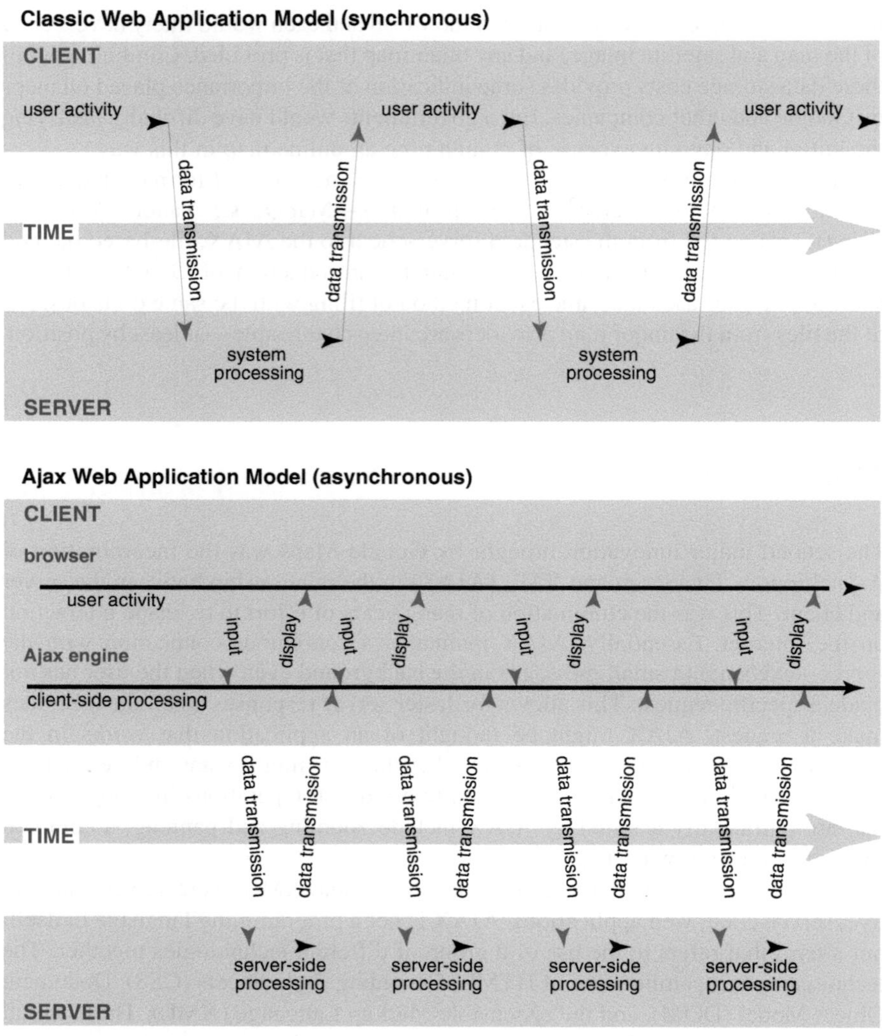


**Fig. 13.5** The typical client–server communication is synchronous (top illustration). AJAX uses asynchronous communication between the client and the server. A connection is maintained to the server to speed interaction

the start of a web session, the AJAX engine is initially loaded in the background. Once loaded, the JavaScript code downloads data from the server without refreshing the web page. A user action that normally would generate an HTTP request to the server becomes instead a JavaScript call to the AJAX engine. If the engine can respond to a user action, no response from the server is required. If the AJAX engine needs something from the server to respond to a user request—such as retrieving new data—the engine makes the request without interrupting the user's interaction with the application. AJAX has transformed the online client/server experience and is used by many different types of applications.

### 13.3.3 Mashups and the Application Programmer Interface

Mashups are an integral part of what is commonly referred to as Web 2.0 (Batty et al. 2010). Beginning in about 2004, Web 2.0 represents a variety of innovative resources, and ways of interacting with, or combining web content. In addition to mashups, Web 2.0 also includes the concept of wikis, such as Wikipedia, blog pages, podcasts, RSS feeds, and AJAX. Social networking sites like MySpace and Facebook are also seen as Web 2.0 applications.

Central to mashups are Application Programming Interfaces (APIs), online libraries of functions that that are made freely available. Many different API libraries have been written for the user-driven web. APIs are the tools that facilitate the melding of data and resources from multiple web resources by providing tools to acquire, manipulate and display information from a variety of sources. In a strict sense, map mashups combine data from one website and display it with a mapping API. The term has come to be used for any mapping of data using an API, even data supplied by the user.

APIs are used for many different types of applications but the creation of maps is one of the major uses. This should not be surprising because there is a great deal of data that has a location component. The relative ease of overlaying all types of information with online mapping tools has further transformed cartography from a passive to an active enterprise with all types of data being mapped.

Introduced in 2006, the Google Maps API consists of a series of map-related functions. These functions control the appearance of the map, including the scale, position, and any added information in the form of points, lines or areas. The purpose of the API is to make it possible to incorporate user-defined maps on websites, and to overlay information from other sources.

### 13.3.4 Map Layers

Google and the other online map providers have a large number of pre-defined layers. These layers have also been tiled—like all of the tiles that make up the map and the satellite image. Table 13.2 shows the available layers provided by Google.

These overlays consist of a series of tiles that have the same size and dimension as the base tiles. Most of the overlay tiles are made transparent so that you can see the tile underneath. Whatever part of the tile is opaque becomes superimposed on the underlying map (see Fig. 13.6). Overlaying transparent tiles in this way is faster than overlaying individual points or lines. Most maps supplied by these services are

**Table 13.2** Standard layers provided by Google. All layers are tiled

| Traffic | Current traffic conditions |
|---|---|
| Photos | Locations of available photographs |
| Labels | Street, city, and boundary names |
| Webcams | Locations of cameras with live imagery |
| Videos | Locations of YouTube videos |
| Wikipedia | Wikipedia pages for locations on map |
| Bicycling | Biking paths and trails |
| Buzz | Postings to Google Buzz |
| 45° | 45° degree angle bird's-eye view |
| Terrain | Shaded relief map |
| Transit | Public transportation network |

**Fig. 13.6** The tile overlay method. A map tile on the left is overlaid with a transparent PNG file with an opaque line. Combining the two tiles produces the display on the right

transparent tile overlays and have a particular theme such as traffic, webcams, or photos.

Most of the layers that are offered by online mapping services are updated only occasionally. An example of a frequently updated layer is traffic (see top map in Fig. 13.7). This layer shows the speed of traffic for major streets in the larger cities of the world and is updated continuously throughout the day. The maps are also tiled at multiple levels of detail for faster download. Only those tiles are replaced that need updating. The bottom map in Fig. 13.7 shows a bicycle path map that completely covers the underlying street map, at least in urban areas. It is updated only irregularly.

An example application of the overlay is the campus map for the University of Nebraska at Omaha (see Fig. 13.8). Here, a map of the campus was first constructed with vectors in Adobe Illustrator. Subsequently, the map was converted to the raster PNG format and tiled through an online tiling service. Three different levels of detail of tiles were created and combined with the Google Map as a layer. Layers were also created for parking and the shuttle transportation network.

Figure 13.9 shows how the parking and shuttle layers are integrated within Google Maps to provide to provide an overall view of the campus. All Google Map functions are active, such as searching for a specific location or travel routing.

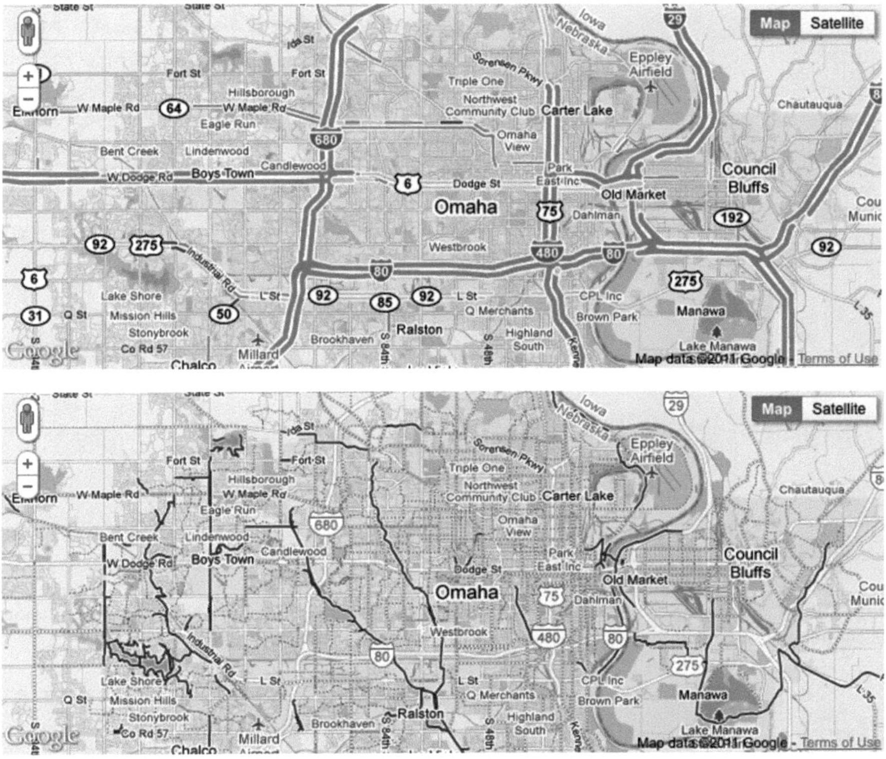


**Fig. 13.7** Comparison of two different overlays. The *top* map showing traffic is transparent allowing the underlying street map to be visible. The *bottom* map showing bicycle paths is opaque, at least in urban areas. Both overlays are supplied in multiple levels of detail

## 13.4 Summary

The advantage of using a major online mapping site is that the maps represent a common and recognizable representation of the world. Overlaying features on top of these maps provides a frame of reference for the map user. A particular advantage for thematic mapping is the ability to spatially reference thematic data. For simplicity, thematic maps have limited the display of spatial reference data such as the location of cities and transportation networks. This has made it more difficult for map users to spatially reference thematic data.

Storing the online map represents a major expense for online map providers. Depending upon the method of storage, the cost for simply storing the map may be as much US $630 million. Maintenance of the database and a variety of other costs would represent significant on-going costs. A non-profit agency or government

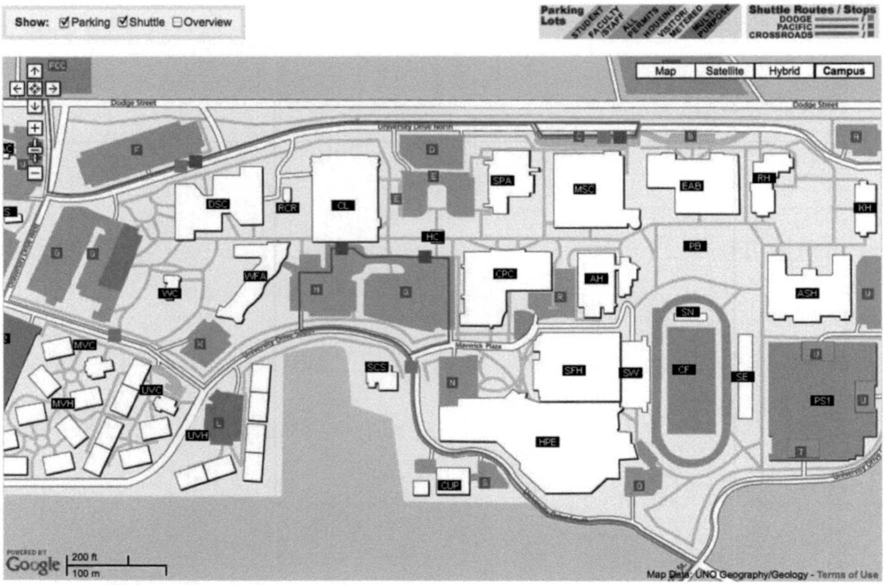


**Fig. 13.8** Campus map overlay for the University of Nebraska at Omaha. The tiled representation is supplied at three levels of detail and overlays over a standard Google Map

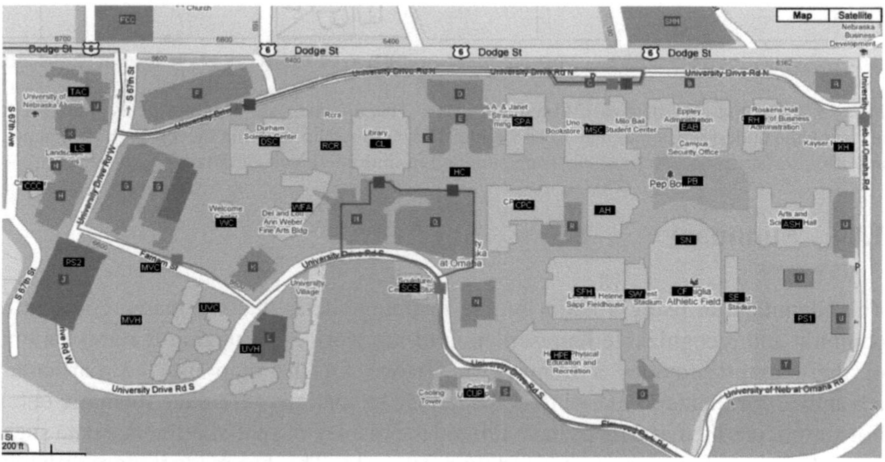


**Fig. 13.9** Parking and shuttle layers superimposed on a Google Map of the University of Nebraska at Omaha campus

entity would have difficulty justifying this expenditure. Application Programmer Interfaces represent a way to integrate information with the online map. For a variety of reasons, this form of map delivery is the most cost-effective way of making maps available to the online map user.

## References

Batty M et al (2010) Map mashups, web 2.0 and the GIS revolution. Ann Gis 16:1–13
Peterson MP (2007) International perspectives on maps and the internet. Springer, Berlin
Peterson MP (2003) Maps and the internet. Elsevier, Amsterdam
Putz S (1994) Interactive information services using world-wide web hyper-text. Comput Networks ISDN 27:273–280
Sample JT, Elias I (2010) Tile-based geospatial information systems: principles and practices. Springer, New York

# Chapter 14
# Visualisation of Geological Observations on Web 2.0 Based Maps

**Gáspár Albert, Gábor Csillag, László Fodor, and László Zentai**

**Abstract** The method of geological mapping has changed during the last decades and the collected data have been recently stored in the records of databases instead of hand-written notebooks. In the Geological Institute of Hungary, a special database structure was designed for primarily scientific purposes, but also for storing and classifying the geological observations according to their importance for geotourism. The relational database of the geological observations can be queried by different subjects and transcribed into KML files, which are useful for the dissemination of geological data via web 2.0 map applications like Google Earth.

## 14.1 Introduction

Before the digital era, the methods and main objectives of a geological description during mapping were worked out in details by several authors (e.g. Compton 1985; Barnes 1995). These methods were both restricted and unobstructed at the same time. Restrictions pre-defined those criteria that were indispensable for the understanding and for the localisation, but the details and lengths of the description were up-to the documenter.

Working out the requirements of a database, which would store the field observations, is usually problematic. The problems originate from the contradiction between the logic of the traditional documentation system and the uniform methods of the technical processing of the data. People, who want to use a common database, think differently, and the database would be usable for them only in the case the logic of the querying methods corresponds with their own logic.

G. Albert (✉) • G. Csillag • L. Fodor
Geological Institute of Hungary, Budapest, Hungary
e-mail: gaspar.albert@gmail.com

L. Zentai
Department of Cartography and Geoinformatics, Eötvös Loránd University, Budapest, Hungary

L. Zentai and J.R. Nunez (eds.), *Maps for the Future*, Lecture Notes in Geoinformation and Cartography 5,
DOI 10.1007/978-3-642-19522-8_14, 

Creators of geological databases worked out different methods to reduce the conflict between the two different logic (e.g.: Laxton and Becken 1996; Brodaric 2004; Clegg et al. 2006; Dey and Ghosh 2008). All of these contributors attempted to collect and store field observation data, however, the developed methods and the structure of the database were different, because the mapping conditions and the priorities were also different.

The field observations are the primary data that make the compilation of geological maps possible for the geologists. These observation data are mostly surveyed traditionally, with paper maps and hand-written notebooks (Fig. 14.1), and stored in archives for the geological survey organizations. The techniques of the digital geological mapping (DGM) developed only during the latest decades. Although several author discussed the methodology and applications of DGM (e.g.: Struik et al. 1991; Briner et al. 1999; Kramer 2000; Jones et al. 2004; McCaffrey et al. 2005), international standards still do not exist for this technology.

Some of the observed sites have not only scientific, but also attractive value. It is obvious for the friends of the natural heritage of the world that many picturesque and visited landforms mostly have geological significance. For the geo tourists, these spot-like points of interest are best known as *geosites* (synonym for *geotops*). Hungary is also full of geological curiosities. Well known geosites, like the Baradla Cave of the Aggtelek and the Slovak Karst, which was inscribed on the UNESCO World Heritage List in 1995 (WHC 1995), have wide publicity, but the minor sites are mostly unknown except for the geologist.

The Geological Institute of Hungary have been collecting the documentation of geological sites for more than 140 years. Although the number of documentations may exceed many thousands, only a few dozen of the documented outcrops would attract the attention of the geo tourists. These sites usually have both scientific and attractive value and thus deserve the name, "geotop".

How can one access these observation data? It became obvious during the last decades that the most effective way of accessing the archives of the old manuscripts is the digitization of them. Therefore, the digitization of the manuscripts of the documentation was one of the aims of the authors' work.

The selection of possible geotops from the swarm of data was another aim. The method that was worked out for these aims comprised two main phases: in the first phase the structure of a relational database was worked out, which would contain the archive documentation of the geological observations. In the second phase, the process of visualisation and dissemination of the selected outcrops was carried out.

## 14.2 Methods

The way of geological documentation in the field was worked out by generations of geologists. The following method can be considered as a novelty only because it represents a new data structure, which is suitable either for reconstructing and digitalizing archive data, or for processing new observations into a relational

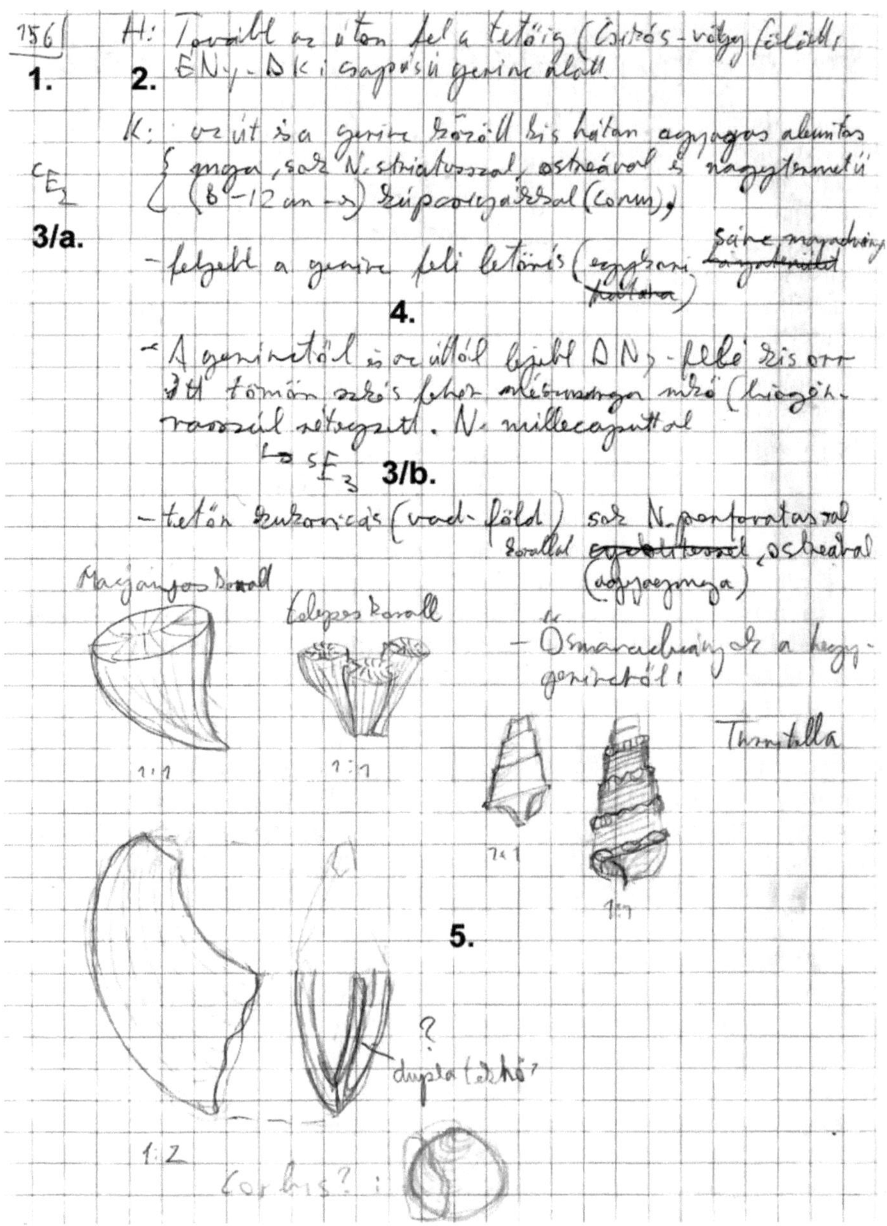


**Fig. 14.1** Hand-written geological observation. 1 = Number of the observation point; 2 = Description of the location; 3/a and 3/b = Geological indexes; 4 = Description of the geological formations; 5 = Drawings and list of fossils (After the notebook of G. Albert 2005)

database. Furthermore, it is applicable for digital field instruments such as PDA (Personal Digital Assistant) or tablet PC.

The experimental application of this method was carried out during the current Regional Mapping Project of the Geological Institute of Hungary in the Gerecse Hills (Northern Central Hungary). The aim was to work out and "fine tune" the general data processing method and the data structure of the observations. It was a basic requirement that the data structure could be processed and queried, but the recorded data should be explanatory and not restricted by the recording methods. It was also necessary to prepare the data structure to receive data from handwritten manuscripts either as typed in as ASCII texts, or scanned as raster images. The data structure was designed to satisfy three functionalities:

1. Recording newly observed but "traditionally" documented data in a structured database (in office);
2. Rapid digitization of archive data (in office);
3. Recording new observations directly in the database (in the field).

The first approach of the data digitization in the Geological Institute of Hungary used the MS Office Applications in the 1990s. Technicians used Excel and Word for recording and typing the handwritten manuscripts into files. It became a routine; however, the data structure changed many times. The firstly created files, especially the Word documents often contained only the explanatory data and the formatting of the documents was not suitable for executing data-mining processes (Fig. 14.2). Therefore, it was necessary to rearrange the data in these documents.

Experience showed that the reshaping (Fig. 14.3) was most effective when the data remained in the same form (in Excel sheet or in Word document) until they could be data-mined and reconstructed in a database. It was also reassuring for the technicians who could work in a familiar environ. To carry out the most effective data-mining, a series of Visual Basic macros was developed to be executed on MS Office documents. These macros had the following functionalities:

1. $p_2^m$

North from the church of Vértesszőlős village and from the entrance of the museum, freshwater limestone was exposed in the abandoned quarry. The famous fossil (dated to 350-500 y.b.p.) of the skull fragment of "Sam" the Neanderthal man was collected from here. The memorial plaque of the finder, László Vértes is placed on a 5 m high, and 20x10 m wide mound in the middle of the quarry yard. Here the limestone is lamellar, slightly curving, with a sequence of 1–10 cm thick beds on the upper 1.2 m (A). Below, 2 m massive bedded, poorly cemented limestone can be found with quite many cavernous plant stem imprints in it (B). The lamellar limestone is friable along the plains of the beds, but consistent and yellowy-brown inside. Bended secessions can also be found locally.

Age: Middle Pleistocene

**Fig. 14.2** Archive digital documentation of an observation from 1991 in MS Word format. The capital letters (A and B) indicate parts of a hand-drawn figure. Note that neither coordinates, nor dates were recorded in the digitized description

| | |
|---|---|
| ID:................................ | LAB3-AG051 |
| Geo Index: .................... | dT3, pJ1 |
| Observer: ...................... | Gáspár Albert |
| Date: ............................ | 1 August 2002 |
| Coordinate: .................. | 611568, 260336 |
| Latitude:........................ | 47.68572222 N |
| Longitude:..................... | 18.53552778 E |
| Elevation:...................... | 263 m |
| Type of outcrop: ........... | in situ |
| Importance:................... | high (Jurassic limestone and cave sediments) |
| Location:....................... | Cliff on a small spur on the western slope, in the curve of the Masina Valley |
| Lithology: ..................... | The 3–5 m rock face consists of massive, thickly-bedded, micritic, Upper Triassic Dachstein limestone with lopher-cycles (dT3). The general dip direction of the limestone is towards the Northeast. There is a plenty of dripstone fragments in the debris around the cliff. Thinly bedded, Lower Jurassic Pisznice limestone, with bulbous bed surface, ammonites and crinoids (pJ1) can also be found in debris on the eastern slope of the valley. There are also many rock debris from the Jurassic "ammonitico rosso" facies. |
| Stratigraphy: ................. | dT3\|pJ1 |
| Measurements:............. | 285/25 bed (dT3) |
| Fossils:......................... | ammonites, crinoids |
| Photos:......................... | drawings in the notebook; 4321-25 |
| Samples: ...................... | LAB3-AG051/1 (ammonite) |
| Notes:........................... | The presence of dripstone debris suggests that there was once a cavern here, which may have collapsed. It is very likely that the Masina Valley was once the cave itself. The Jurassic rock debris may fell into the cave through a sinkhole, but it is also possible that the Jurassic limestone was in situ nearby or tectonically transported to the present location. Eocene gravels and debris were not observed. |
| Alias: ........................... | |

**Fig. 14.3** General geological documentation of an outcrop in the Gerecse Hills (Northern Central Hungary) in the pre-defined MS Word format, which was designed to easily execute data-mining processes on it

1. Searching for data in Word documents and exporting them into a database;
2. Searching for missing or corrupt data fields in the created database;
3. Decoding data-fields using uniform names;
4. Correcting mistyping in selected data-fields;
5. Exporting the selected attributes into XML files.

The result of the data-mining and processing was a database of the geological observations, which was stored in MS Access tables. The XML (Extensible Markup

Language) files were created for data exchange purposes. The following section gives a detailed description of the re-shaped document structure, which can be considered as fields or collections (queries) of the final database.

### *14.2.1 Database of the Geological Observations*

The geologist, who is responsible for the documentation, uses numbers and letters to create a code for each registered outcrop during mapping. It is the way of documentation now and it was the same a hundred years ago. During these many years, the codes naturally change and one code may often occur repeatedly, thus in creating the new database it was a primary requirement to work out a unique and easily understandable naming method for the database entries.

The mapping is carried out at large scale (1:10,000) state topographic maps. The projection of these maps is the official Hungarian National Grid (EOV) system, but the coordinates can be recorded or transformed into latitudes and longitudes on standard WGS84 datum.

The database is represented by several main, sub- and code tables (Fig. 14.4). The main tables contain genuine data, like the observation points themselves, or the maps of the region concerned. Sub-tables contain details of the observations, like the measurements and photos taken, or the samples and fossils collected. The code tables store the repeatedly occurring data, such as the objects measured (e.g. fault plane), the names of the fossils, and the observers.

Each record in the "observations" main data table is identified by a unique key in the "op_id" field, which links this table with other sub-tables in the database. It was considered that the database would be better understandable for the geologists, if the unique identifiers are designed to contain real names and numbers. These kinds of identifiers are called natural keys. To produce a natural key field, a 4-letter code of the documented area was put together with the monogram of the documenter and the number of the observation point (e.g.: LAB3-AG051 is the natural key for the 51st site of Gáspár Albert on the map, which is coded as LAB3). The area codes are also natural keys, since they are abbreviations of the name of the topographic maps. For example, the code LAB3 is the third sub-sheet of the map called "Lábatlan", named after a little village in the area.

The rules of the documentation are strict. From the formatting of the identification numbers to the applicable characters in the string data of the geological descriptions, are all strictly defined. Without the restricting VBA (Visual Basic Applications) codes, hardly a few documenters would keep themselves to the formalities. These codes (macros) are built in the model Word document, and they also handle the link to the database.

The field documentation is slightly different, since these codes are not compatible with the mobile applications of the data recording device. Thus the data, which were recorded in the field, go through a syntax-controller program before the processing.

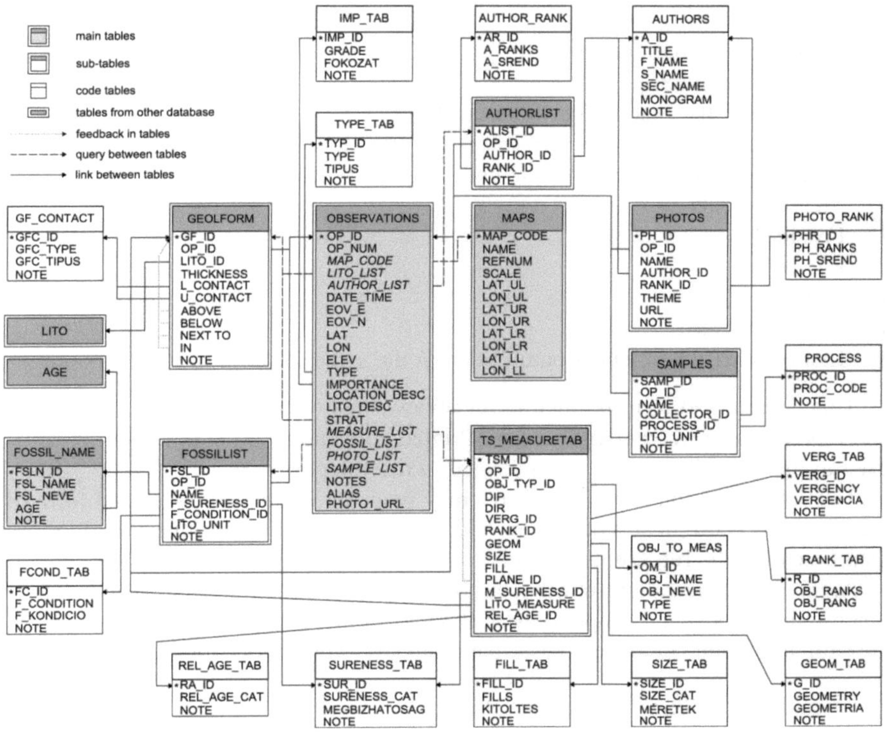


**Fig. 14.4** Structure of the relational database of the geological field observations. The code tables are bilingual (English and Hungarian), the language of the documentation is selectable. Field names in italics in the OBSERVATIONS main data table indicate that the data recording process may happen directly or through the linked sub-tables

## *14.2.2 Data Fields or Categories?*

If the data of the observation are collected in one form (Fig. 14.3), each category represents one ore more records in the data tables. These forms are originally the Word documents themselves, created by the documenters, but the creation of these collections may work reversely if it is needed.

Creating the collections from the database is a question of querying the database. In a query, which produces the form of the Word document (Fig. 14.3), the records are collected from the main or sub-tables of the relational database.

Both in the forward and in the reverse process, the categories are basically the followings:

1. Identification of the outcrop
2. List of the identified objects (with codes)
3. List of the documenters
4. Date and time of the documentation

5. Coordinates
6. Importance of the outcrop
7. Type of the outcrop
8. Location of the outcrop (with description)
9. Geological description of the observed formations
10. Stratigraphy and relation of the observed and identified objects
11. Measurements
12. Fossils
13. Photos
14. Samples
15. General notes
16. List of other documented outcrops from the same location or nearby

This categorized form is suitable for reviews and overviews of the observation, and represents a printer friendly form. Printed collections of the outcrops are often used in the fieldtrips and as appendices of written reports.

The categorization described above may reappear in visually enhanced form of the observed geological data, but in a visual representation of a query, only those data fields will be represented which have a value (Fig. 14.5).

Archive documentations rather focus on the colour, the texture, and the type of the outcropping rock (e.g. Compton 1985; Barnes 1995). Although these characteristics are still important, field geologists have already worked out methods to classify

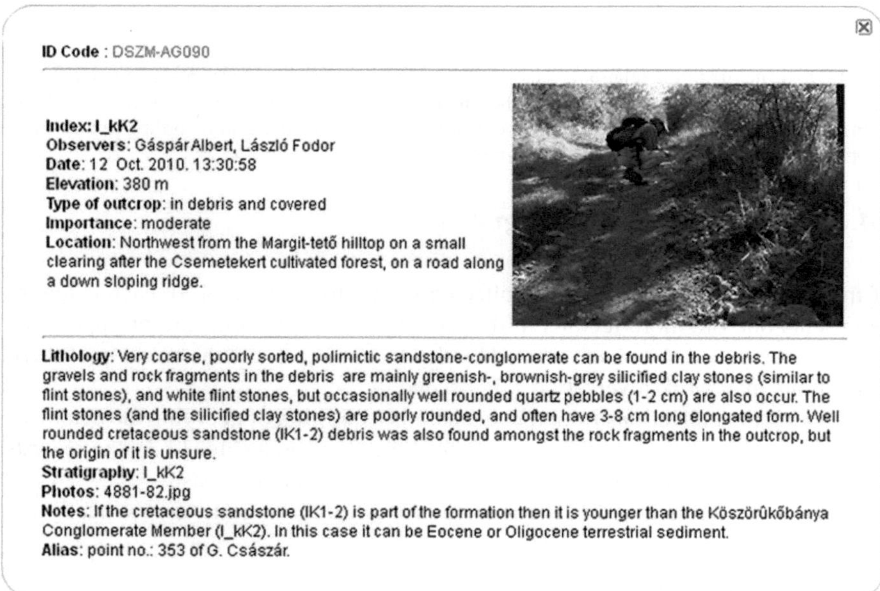


**Fig. 14.5** Visually enhanced representation of a geological observation as a Google Earth "bubble". The categories of the documentation were queried from the database of the observations and were transcribed into XML files

sedimentological, tectonic, volcanological, etc. features in the field, which help to determine the genetics of the observed rock. These features can be recorded in the field both in DGM devices, and in traditional notebooks, but finally the data within the database will be stored in sub-tables.

### 14.2.3 Visualisation of the Observed Data on XML Bases

The XML (Extensible Markup Language) was created to store documents in machine-readable form. It was developed by the W3C (World Wide Web Consortium) in 1998 from an ISO standard (SGML = Standard Generalized Markup Language) with the purpose to store and transfer data via Internet (ISO 8879 1986; W3C XML 1998). The structure of an XML file is defined in a schema document (XSD, or XMS), which, like an empty database, contains information about the data-fields and types. The KML files, officially named the OpenGIS® KML Encoding Standard, are also XML documents, but specialized for the description of geospatial data. Since both the XML and the KML are open standards, data collecting, processing and visualizing applications can be developed freely to imply these data structures. The KML is maintained by the Open Geospatial Consortium Inc. (OGC 2008), and was improved several times since its first publication. The KML standard became widespread along the expansion of Google Maps and Google Earth.

The structure and the content of the geological observation database were translated into KML files to create Google Earth compatible selections of specific outcrops. In the process, the "ogckml22.xsd" XML schema file was used. The criteria of the selection varied according to the purpose of the visualisation. Basically, the following criteria were used:

1. Outcrops of public interest
2. Outcrops of a specific area
3. Outcrops documented by a specific person

The first category was restricted to those outcrops which had high importance. Some of the KML files were used by the geologists of the Regional Mapping Project, some of them were published on the Internet to disseminate the geological knowledge (Albert and Csillag 2010).

Besides a wide range of documents, the KML format can also describe georeferenced images, spatial vectors, polygons and shapes. This allows one to transcribe maps in raster or vector file format into KML files. It is especially useful if the user interface—preferably a free software package for visual representation of the data—works with this kind of file structure. Thus, the documented geological data might be visualized together with the photos and geological maps (Fig. 14.6).

The transcription of the geological maps of selected regions was carried out with the Global Mapper 10 software. For the user interface the Google Earth 5.2 (GE) freely downloadable software was applied (Google Earth 2010), but any GIS

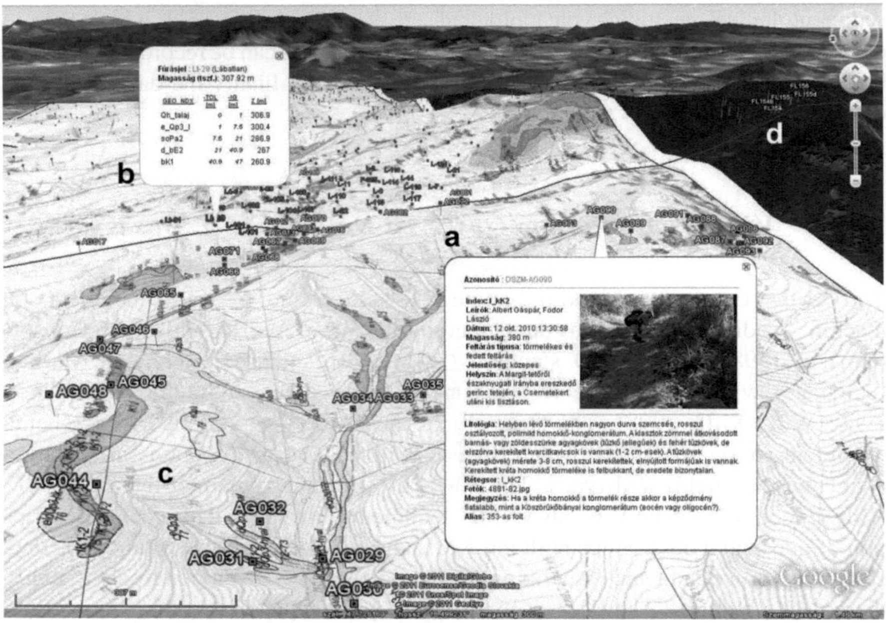


**Fig. 14.6** Screenshot montage of the GE interface showing: (**a**) The data and photo of an observation point (*lower right*), (**b**) a borehole log from the borehole database (*upper left*), (**c**) a geological map draped on the DTM, and (**d**) the boundary of the mapping areas (*dark lines*). These data together enable one to easily overview the spatial relations between the geological formations

software, which is capable to read the KML format, is suitable for this purpose. The free tool was selected for the visualization, because an unlimited number of users can access the files after creation and interpretation of the data-packages. Furthermore, the visual representation of the geological data in GE is enhanced with built-in background of satellite images, digital terrain models and a wide range of arbitrarily selectable themes. The following maps were displayed for geo-scientific purpose along with the observational database:

- Geological maps covered and uncovered by Quaternary sediments
- Geochemical maps
- Topographic maps
- Geophysical (e.g.: Bouguer anomaly) maps
- Geological maps for tourists
- Hydro-geological maps

The geo-scientific approach of the visualization of geological data brings up the necessity to visualize not only the observations, but, for example, the boreholes in a given area too (Fig. 14.6). The borehole database as well as the observation database can be transcribed in KML format. For the querying of a database, different from the observations, a Visual Basic script was programmed. With the help of the script, the transcription of the query in KML files was carried out automatically from the selection of data fields, and the data were visualized ergonomically in the GE interface.

## 14.3 Results

The advantages of a relational database and the freedom of a traditional handwritten documentation were combined in the newly developed documentation system. This duality is represented in the authors' documentation methods, and in the structure of the database.

The number of photographs of an observed site has rapidly increased since the times of digital photography. These photos themselves formed quite large digital data, which were necessary to categorize and store in some kind of database. The expansion of the GPS usage in the field also led to the creation of sets of data. Although the created database structure follows the logic of the field notes, both the usage of the digital photography and the GPS represent such a technical background that is suitable for a connection between these data.

The method of the dissemination of the collected geological knowledge was the other achievement besides the construction of the geological database.

The increasing attempt to elongate the season for tourism was one of the main reasons for the development of the geo-tourism. Visiting geological outcrops practically needs only good weather conditions, and except for those latitudes where snow covers the geological formations in winter, any season would be suitable for this purpose. The establishment of new national parks and nature reserves and their widening profile give new challenges for the geoscientist, because the natural heritage in the concerned regions consists not only of the flora and fauna, but the geological background as well.

A national survey of geosciences has to deal with this new challenge and find ways to disseminate useful knowledge of the geologically interesting and scenic locations. These geosites may be located on already preserved lands, but often they are not. Abandoned quarries, collapsed caves on a construction site, or picturesque landforms in private forests are often the subjects of such curiosity. Some of them may be preserved only as memories, because the quarry would be recultivated, and the cave would be buried or covered on the construction site. These memories are the documentations and photos taken by the geologists.

The new database structure made it possible to classify the geological objects according to their significance whether they were recorded newly or data-mined from decades-old archives. The applied method for the visualisation made it possible to put selected sites (geosites) and maps on the Internet for public use. These two achievements are just a small part of the work, which needs to be done.

The legislation of the natural heritage is usually different in each country, but the non-living part of our natural environ is considered a subject of protection only recently even in the environmentally aware countries. On the long run, it is inevitable for the geological surveys to mark the reference locations of all known geologically interesting landforms and to select those that need legal protection to be preserved.

## 14.4 Discussion

Before and partly parallel to the development of the geological database and documenting system, some previous methods were applied and tried in various geological mapping projects. These methods were developed mainly by programmers, and the aim of the documentation was different. With the collaboration of the Geological Institute of Hungary, a research project was initiated. Exploration wells were drilled in order to find a safe location for the low and intermediate level radioactive wastes in Hungary (Balla 2004). For the on-site borehole documentation, handheld PDA and an XML-based data collecting application was used in this project (Gyalog et al. 2004).

Although at that time the XML format was quite modern, the database structure was rigid, and when the documentation was over, the data were not easily accessible for the geologists. Much of the effort spent on the input of the observed data was futile. On the other hand, a quite usable, user interface was developed to query the borehole data in an Internet browser (Gyalog et al. 2005). This interface was designed to query only a thin segment of the overall database, but this part—a simplified stratigraphic column of each well—was the most significant in some way (e.g.: enabled the correlation between the borehole data and the medium scale geological maps).

Despite the deficiencies of the results, it was an important step to apply and try preliminary documentation systems and user interfaces. It is futile to believe that a database structure for "all use" can be created. The database can be considered perfect when the stored data can be accessed easily and there are no redundancies in it. The users of a geological database also vary. Basin modellers search for the lithological data, while the palaeontologists are interested in the fossils.

The nature of the geological field observation is quite unique. The range of observation is usually limited because of time, weather and terrain conditions. The details of the described geological formations are limited to the eye and loupe range, and the number of collected samples is often limited by the physical condition of the collector.

The edification in the previous borehole documentation and querying system was that the documenters should not spend their time picking data from pre-defined lists and fill data fields which are not or hardly accessible later. The time consuming on-site documentation will lead to the simplification of the geological description, and thus to a decreasing scientific value. The pre-defined—often very long—drop-down lists on a data recording interface can also lead to the same result.

The digital representation of the database is usually only a question of the GIS background. If the financial conditions allow it, one can use several kinds of software, which not only create and manage, but also properly visualize the database background and the thematic maps of a region. However, this kind of complex GIS functionality is quite expensive and usually only a limited number of licences are affordable for a geological survey. The visualization of data on the web 2.0 using free licensed software solves the problem of the licence limitation and makes it possible to publish geological information freely and properly.

Publications of point-of-interests, satellite images and maps draped on a 3D surface serve as useful tools to disseminate scientific knowledge amongst those, who are interested, but not specialized in geology. The authors' experience shows that the map applications of the web 2.0 are useful tools for geoscientists as well.

**Acknowledgments** The authors thankfully appreciate the efforts of the documenters who took care of using the prototypes of the database, and occasionally made suggestions and gave advice. An ergonomic documentation system and a well-structured database serve the benefit of not only the geologists of the Geological Institute of Hungary. On the long run, the base application can be widely used amongst the Hungarian and maybe the foreign earth scientists too.

## References

Albert G, Csillag G (2010) Földtani tartalomszolgáltatás web 2.0 alapokon—Geological knowledge service on web 2.0 base. http://hantken.mafi.hu/pub/rastermaps/Albert G, Csillag G 2010 Foldtani tartalomszolgatatas web 2.0 alapokon.pdf. Accessed 11 Aug 2011

Balla Z (2004) General characteristics of the Bátaapáti (Üveghuta) site (South-western Hungary). Ann Rep Geol Inst Hung 2003:73–91

Barnes JW (1995) Basic geological mapping, 3rd edn. Wiley, Toronto

Briner AP, Kronenberg H, Mazurek M, Horn H, Engi M, Peters T (1999) FieldBook and GeoDatabase—tools for field data acquisition and analysis. Comput Geosci 25:1101–1111

Brodaric B (2004) The design of GSC Fieldlog: ontology-based software for computer aided geological mapping. Comput Geosci 30:5–20

Clegg P, Bruciatelli L, Domingos F, Jones RR, De Donatis M, Wilson RW (2006) Digital geological mapping with tablet PC and PDA: a comparison. Comput Geosci 32:1682–1698

Compton RR (1985) Geology in the field. Wiley, New York

Dey S, Ghosh P (2008) GRDM—a digital field-mapping tool for management and analysis of field geological data. Comput Geosci 34:464–478

Google Earth (2010) version 5.2.1.1588. http://www.google.com/earth/index.html. Accessed 1 Sept 2010

Gyalog L, Havas G, Maigut V, Maros GY, Szebényi G (2004) Geological-tectonic documentation in the Bátaapáti (Üveghuta) site. Ann Rep Geol Inst Hung 2003:171–201

Gyalog L, Orosz L, Sipos A, Turczi G (2005) The uniform legend system, the borehole database and the web-based query tool of them in the geological institute of Hungary (in Hungarian, with translated abstract, and figures). Ann Rep Geol Inst Hung 2004:109–125

ISO 8879 (1986) Information processing—text and office systems—Standard Generalized Markup Language (SGML). http://www.iso.org/iso/catalogue_detail.htm?csnumber=16387.Accessed 4 May 2011

Jones RR, McCaffrey KJW, Wilson RW, Holdsworth RE (2004) Digital field acquisition: towards increased quantification of uncertainty during geological mapping. In: Curtis A, Wood R (eds) Geological prior information, vol 239. Geol Soc Spec Publ, London, pp 43–56

Kramer JH (2000) Digital mapping systems for field data. In: Soller DR (ed) Digital mapping techniques '00-Workshop Proceedings. US Geol. Survey Open-File Rep 00-325:13–19. http://pubs.usgs.gov/openfile/of00-325/kramer.html. Accessed 2 May 2011

Laxton JL, Becken K (1996) The design and implementation of a spatial database for the production of geological maps. Comput Geosci 22:213–225

McCaffrey KJW, Jones RR, Holdsworth RE, Wilson RW, Clegg P, Imber J, Holliman N, Trinks I (2005) Unlocking the spatial dimension: digital technologies and the future of geoscience fieldwork. J Geol Soc London 162:927–938

OGC (2008) XML Schema Document for OGC KML version 2.2. http://schemas.opengis.net/kml/2.2.0/ogckml22.xsd. Accessed 4 May 2011

Struik LC, Atrens A, Haynes A, (1991) Handheld computer as a field notebook, and its integration with the Ontario geological survey's "Fieldlog" program. In: Current research, Part A, Cordillera and pacific margin. Geol Surv Can 91-01A:279–284

W3C XML (1998) World Wide Web Consortium—XML Technology. http://www.w3.org/standards/xml/. Accessed 4 May 2011

WHC (1995) UNESCO World Heritage Convention—Caves of Aggtelek Karst and Slovak Karst. http://whc.unesco.org/en/list/725. Accessed 4 May 2011

# Chapter 15
# Updating a Hungarian Website About Maps for Children

**José Jesús Reyes Nuñez and Csaba Szabó**

**Abstract** The first website in Hungarian language dedicated to the presentation of basic cartographic concepts for children and young people was made in 2000, preceded by a study about the basic map concepts learnt by Hungarian pupils in Elementary and Secondary Schools. After more than 10 years from the publication of the original website, the time has arrived to update the website, considering that Internet and in particular the Web has had notable changes along this period of time and also considering possible changes in the teaching of the map concepts in the Hungarian Educational System. This decision implied not only the possible revision of the themes presented in the original website, but also the analysis of compulsory changes in the structure of the website, namely, which new programming and design tools can be used to develop the new site and which new graphic tools can be added to develop a new Web based environment.

## 15.1 Original Research (1997–2000)

### *15.1.1 Theoretical Research*

The project was begun with the study of curricula, textbooks, workbooks and atlases used in the subjects related to Geography in the Hungarian Educational System, to determine which cartographic concepts were learnt by the pupils. The conclusion was that the majority of the cartographic concepts were taught between the third and the fifth grade of Elementary School (Tompané 2002), and the posterior use of maps and school atlases in the classroom (including High School) is based on this knowledge (Horváth et al. 2002a, b).

J.J.R. Nuñez (✉) • C. Szabó
Department of Cartography and Geoinformatics, Eötvös Loránd University, Budapest, Hungary
e-mail: jesus@ludens.elte.hu; jesus@map.elte.hu; bahama@map.elte.hu

L. Zentai and J.R. Nunez (eds.), *Maps for the Future*,
Lecture Notes in Geoinformation and Cartography 5,
DOI 10.1007/978-3-642-19522-8_15, 

Parallel to this research a study was also made about different methods of publishing educational materials on the Web, paying special attention to finding international experiences in the presentation of concepts related to maps. In the last years of the 1990s, the number of homepages designed to fill this aim could be considered still low. By this reason the research was extended to homepages designed for children in interest of presenting concepts related to different sciences.

The next step was to classify the map concepts for their presentation on the Web, first dividing them into general themes, followed by their detailed description, the definition of smaller units and finally completing the themes taught in schools with other concepts (Reyes 2002a, b).

The list of general themes presented in this website was: Map and reality, Orientation with and without maps, Map history, What kind of maps are there?, Representation of relief, Rivers and lakes on maps, Other colours on maps, Symbols on maps, Latitude and longitude, Some words about the geographical names.

### *15.1.2 Planning, Design and Making of the Website*

During the design of the website the decision taken was to follow the structure of a Web portal, not only to present the map concepts but to complete them with other options that could make more interesting the site for pupils and teachers. The title of the website was "All about maps..." and the structure was determined as follows (Fig. 15.1):

- Summing up ...
  Previous works related to this theme.
- As you are learning and even more ...
  Presentation of basic map concepts
- What can you find on the Web?
  Links to websites related to Maps and Children.
- Try out!
  Demos, examples, etc. that were found in other websites.
- News and events.
  Information about activities related to Cartography: organization of exhibitions, Barbara Petchenik Award, meetings, interests on TV programs, etc.
- Not only for you ... but for teachers too!
  Open forum for the exchange of opinions and questions.
- Learn and play!
  Tasks and games related to Cartography.
- Curiosities
  Section including old maps, imaginary or fantasy maps, etc.

The website was programmed using HTML language and some JavaScripts for specific solutions (e.g. opening a separated window to show a short animation).

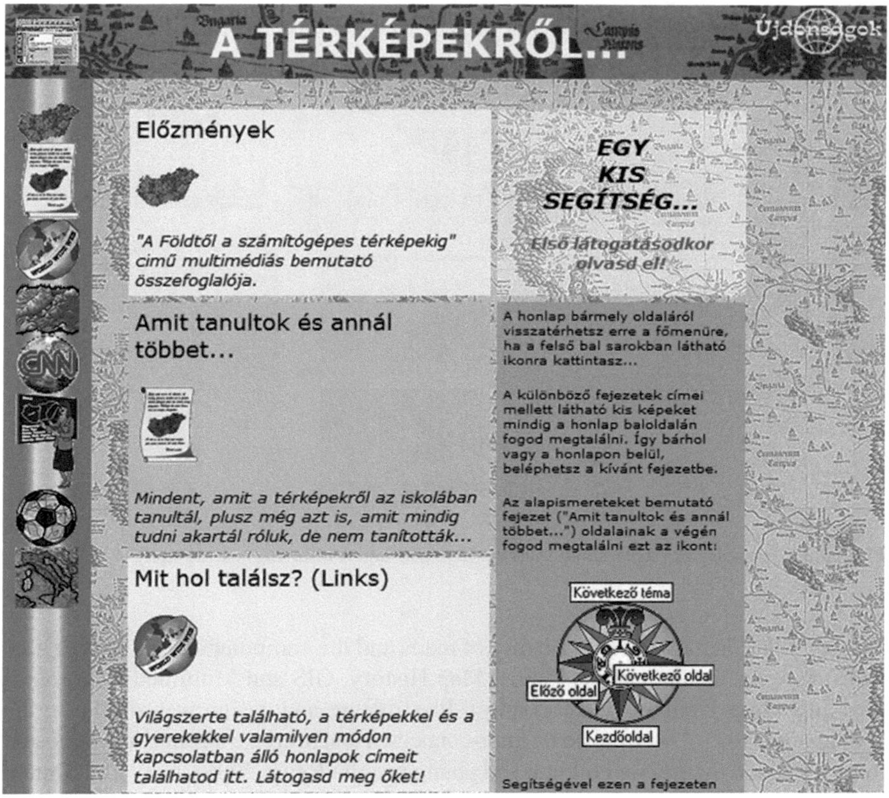


**Fig. 15.1** "All about maps..." website welcome page

After testing it, the website was made accessible for Hungarian pupils and teachers from July 2001 (Reyes 2002).

## 15.2 Update of the Website from 2001 to 2011

During the last 10 years, the structure and design of the website remained changeless, but it was periodically used to keep pupils and teachers informed about activities related to cartography; especially about the annual organization of GIS Days and the organization of the Barbara Petchenik World Map Award every 2 years.

Between 2004 and 2006 new websites were created to complete the content of the chapter dedicated to present basic concepts. This decision was taken considering the results of a survey filled by pupils of Elementary Schools that included a question about their preferred cartographic themes. Based on their answers the two

**Fig. 15.2** Examples of pages about map history, GIS and multimedia cartography

most popular themes were the history of maps and the computer cartography. Three websites were made on these topics: Map History, GIS and Multimedia Cartography (including Thematic Cartography). Its structure and design was similar to the "All about maps..." website, to be inter-connected and to create a collection dedicated to cartographic themes for children and young people (Reyes 2006a, b, c). The content of this collection was composed as shown below (Fig. 15.2).

Map History website (six chapters containing 39 themes):

1. Maps from the Ancient Age: Town Plan from Catal Hüyük, rock map from Bedolina, Mesopotamian world map from Nuzi, Mesopotamian town map of Nippur, Babilonian clay tablet, Turin Papyrus.
2. Hellas and Rome: Pythagoras and the spherical Earth, Aristotle and Dicaearchus, Eratosthenes, Ptolemy, Tabula Peutingeriana, the first T-O maps.
3. Early Middle Ages: Maps from the Middle Ages, monastery maps, climatic zone maps, the Islam Atlas, Al-Idrisi's works, portolan maps, Catalan atlas, translations of Ptolemy's works.
4. The Great Discoveries: Portuguese discoveries, Christopher Columbus, discovery of the New World, Cosa worldmap, Cantino worldmap, Waldseemüller and America, Piri Reis worldmap, Magellan and the circumnavigation of the Earth.
5. The first atlases: Mercator and Ortelius, Gerhard Mercator's works, Abraham Ortelius' works, Waghenaer's atlas.
6. The first Hungarian maps: the English Saxon map, Dulcert's portolan map, Cusanus and Fra Mauro, Lázár secretary's map of Hungary, details on the first Hungarian map, actual Hungarian limits on the Lázár's map, János Zsámboky.

GIS website (six chapters containing 35 themes):

1. About GIS history: What is GIS?, the first steps, analogue GIS, the beginnings of computer sciences, Canada Land Inventory, SYMAP software, ESRI & Intergraph, MapInfo & Autodesk, GIS Day.
2. GIS data: Vector data model, raster data model, layers on the maps, fundamentals of databases, geocoding, multimedia databases.
3. GIS data sources: Surveys, aerial photos, satellite photos, GPS, source maps, thematic databases.
4. GIS processes: data collection, data input, data analysis, data output.
5. Data analysis: Overlay, buffer, classification, fuzzy.
6. Graphic outputs: webmaps, 3D models, animations, Geography Network, Hungarian websites.

Multimedia in maps website (six chapters containing 35 themes):

1. Stories about data visualization: Catalan Atlas, John Graunt, William Playfair, Alexander von Humboldt, making the diagrams popular, Bertin and the graphic theory.
2. The first thematic maps: Halley's map, Valentin Seaman, early geological maps, Humboldt and Berghaus, French influence, John Snow.
3. Thematic maps today: Sciences on maps, drawing symbols, drawing areas, isolines, diagrams on maps, drawing with points, choroplets and cartograms, representation of movements.
4. Traditional multimedia on maps: Before the multimedia of today, what is multimedia?, maps and multimedia, multimedia in atlases, the first animated maps.
5. Digital multimedia on maps: Introduction, beginnings of digital multimedia, digital maps, digital multimedia atlases.
6. Multimedia in webmaps: maps in the Web, interactivity in the Web based atlases, Geographic Network, learning with outline maps, settlements on Lázár's map, "All about maps...".

## 15.3 Creation of a New Website

### *15.3.1 Theoretical Research*

After more than 10 years from the presentation of the original website, the time arrived to update the content if needed, and to modify the structure and design of the website, considering the development experimented by the Internet and in particular by the Web during this period of time. In October of 2010, the authors decided to begin the preliminary tasks to make these changes. First of all, we studied again the content related to Geography subjects in the Hungarian Educational System, paying special attention to the basic level (Elementary Schools). This analysis began with

**Table 15.1** Basic concepts related to cartography in Hungarian school atlases

| My first atlas (Cartographia 2008) | School atlas for 10–16 year-old pupils (Cartographia 2008) | Secondary school atlas (Cartographia 2008) |
|---|---|---|
| 1. **Our world**: rotation and translation of earth, other astronomical concepts | 1. **Representation of relief on a map** : panoramic view, cross-sections, contour lines, hypsometry and relief shading | 1. **Representation of relief on a map**: perspective image, cross-sections, striping, contour lines, hypsometry and relief shading |
| 2. **Map symbols**: legend explaining symbols used in the atlas | 2. **Comparison of maps at different scales**: fragments of maps at 1:25,000, 1:50,000, 1:100,000, 1:500,000, 1:1,250,000 and 1:20,000,000 | 2. **Comparison of maps at different scales**: fragments of maps at 1:25,000, 1:50,000, 1:100,000, 1:500 000, 1:1,250,000 and 1:20,000,000 |
| 3. **From a sketch to a map**: from the school buildings to a large scaled map of the school and surroundings | 3. **Satellite image**: image at 1:500,000 (Budapest and surroundings) | 3. **Satellite image**: image at 1:500,000, Budapest and surroundings |
| 4. **Basic geographical concepts**: main shapes of relief and representation on a map. Landscape and map | 4. **Map symbols**: legend explaining symbols used in the atlas | 4. **Map symbols**: legend explaining symbols used in the atlas |
| 5. **Maps and scales**: panoramic view, tourist map (1:40,000), topographic map (1:100,000), maps at 1:550,000, 1:1,500,000 and 1:20,000,000 | 5. **Cartographic projections**: cylindrical, conic and azimuthal projections | 5. **Cartographic projections**: similar to the atlas for 10–16 years-old pupils |
| 6. **Orientation on a globe and Nature**: determination of cardinal points and geographic coordinate system | 6. **Basic astronomical concepts**: structure of the universe, planets of the solar system, the visible side of the Moon, solar and lunar eclipse, the Earth and the Moon orbit, Earth bound, the Sun's apparent motion, the Earth from the Moon | 6. **Basic astronomical concepts**: similar to the atlas for 10–16 years-old pupils |
| 7. **Methods for the representation of relief on maps**: contour lines, hypsometry and shading | 7. **Types of Hungarian settlements on maps at 1:30,000**: one street settlements, chessboard-structured settlements, conglomeration, circular-structured settlements, settlements connected by bridge, planned industrial settlements | 7. **Types of Hungarian settlements on maps at 1:30,000**: similar to the atlas for 10–16 years-old pupils |
| 8. **Satellite images**: false colour satellite image of Budapest (explanation) | 8. **Tourist map at 1:40,000** (Danube Bend, Legend) | 8. **Maps of landscapes**: estuary, glacier and fjord, agglomeration, agricultural and industrial region on maps |

(continued)

**Table 15.1** (continued)

| My first atlas (Cartographia 2008) | School atlas for 10–16 year-old pupils (Cartographia 2008) | Secondary school atlas (Cartographia 2008) |
|---|---|---|
| **9. Types of settlements**: city, town and farm in an aerial photo and on a map | **9. Other thematic maps:** climate, soils, political, economic, nationalities, ethnographic | **9. Tourist map at 1:40,000** (Danube Bend, Legend) |
| **10. Physical and political maps**: Hungary, Europe and Earth | | **10. Other thematic maps**: similar to the atlas for 10–16 years-old pupils |
| **11. Other thematic maps**: climate, soils, political, economic, nationalities, ethnographic | | |

the reading of textbooks related to Geography to determine if there was any kind of significant changes during the last 10 years (Hartdégenné et al. 2010). It was completed with the study of some of the School Atlases used in the Elementary and Secondary Schools of Hungary (Cartographia 2008a, b, c) to determine the basic cartographic concepts presented or at least mentioned in the introductory pages of these atlases (Table 15.1).

After 2000, the number of hours/week dedicated to the teaching of Geography subjects remained the same: 3 h at basic level (Elementary Schools) and 4 h at secondary level (Secondary and Vocational Schools). In grades 4 and 5, Hungarian pupils learn a subject entitled "Basics about Nature", combining knowledge mainly from Geography and Biology. The concepts related to cartography have not been modified considerably, but in numerous schools the teaching of these concepts has been moved to later grades in respect to our previous research in 1997–2000. The more detailed explanation of these concepts follows being presented in the textbooks written for the fifth grade ("Basics about Nature", see Table 15.2), but some regional curricula planned the teaching of them in grade 6.

After revising the themes presented in the original version of the website and comparing them with the content of textbooks and school atlases, we decided to use the same map concepts without any considerable change. Based on this decision, the structure of chapters remained the same and the detailed content of the website is as follows:

- Map and reality (seven themes: Earth and former worldviews, shape of the Earth, what can we find on a globe, from the globe to a plain: how to make a map?, what is a map?, what makes a map a map?, graphic scale)
- Orientation with and without maps (eight themes: How could a cave man orient?, the first compass, compasses in Europe, cardinal points, using a compass, orienteering with the Sun, in the forest and at night)
- Map history (nine themes: the first maps, maps on rocks, on clay and papyrus, Greek scholars, Roman maps, maps of the Middle Ages, portolans, maps of Discoverers, Mercator and Ortelius, the first Hungarian map)

- What kind of maps are there? (four themes: classification of maps according to their scale and content, thematic maps, city, road and topographic maps)
- Representation of relief (three themes: basic shapes of relief, contour lines and hypsometry, shading)
- Rivers and lakes on maps (two themes: how to draw rivers and lakes on a map?, special features related to hydrography)
- Other colours on maps (five themes: colours on a political map, how to draw the frontiers and roads?, colours on a topographic map, colours on a thematic map, other topics represented on maps)
- Symbols on maps (three themes: what kind of map symbols do you know?, symbols drawing hydrography and relief, symbols in the school atlases)
- Latitude and longitude (four themes: geographic coordinate system, latitude, longitude, orientation on the globe)
- Some words about the geographical names (five themes: what is a geographical name?, grammar of geographical names, using fonts and placing names on a map, hydrographical and relief names, the interest of a better map reading)

At same time, we also kept the themes explained in the websites dedicated to Map History, GIS and Multimedia maps, which were developed between 2004 and 2006.

## *15.3.2 Technological Questions Developing the Website*

After taking the decisions related to the structure of the new website, we faced another not less important question: which environment can be most appropriate to develop the site? Our first choice was to select PHP (Hypertext Pre-processor), which is a widely used general purpose scripting language suited for Web development and which can be inserted into HTML (HyperText Markup Language) too. Searching for other possible options the use of a Content Management System (CMS) was considered a better choice, because this kind of environment facilitates the development of a website offering pre-programmed tools, whose use is easy to learn. The next question was: which CMS should be selected? At present, there is a big number (more than 100) of freely downloadable CMS in the Web: Wordpress, Drupal, Custom CMS, Joomla, Moodle, Typo3, e107, PHPX, PHP SiteManager, PHP SiteManager, etc. Our first condition was to choose a CMS with Hungarian support and the second one was the security that the CMS can offer, preventing our server from possible hacking attacks across the Web. Considering the different options in the market, our decision was to develop the new website in the CMS named Plone.

### 15.3.2.1 What Is Plone? Why Plone?

Plone is an open source content management system (CMS), whose heart is the Zope (Z Object Publishing Environment) application server. Zope is an open

**Table 15.2** Cartographic concepts in the textbooks for fifth grade

| Basics about nature | | | |
|---|---|---|---|
| Title of the chapter/ themes | National publishing house (2002) Our important partner: the map (pp. 104–122) | Mozaik publishing house (2004) Orientation on maps (pp. 34–50) | Apáczai publishing house (2004) Orientation on maps and in nature (pp. 69–84) |
| 1.Map definition | Simplification, reduction, faithful representation | top view, reduction, projection to a plane | Reduction, base plan, top view |
| 2. Scale | Scale | Scale, large and small scale | Scale |
| 3. Graphic scale | Explanation about its use | Explanation about its use | Explanation about its use |
| 4. Earth globe | The reduced image of the Earth | – | – |
| 5.Orientation | Direction, distance, cardinal points, compass, its use and principle (magnetism), Sun based orientation | Cardinal points, compass, its use and principle (magnetism), orientation of a map, orientation with the North Star | Reference system and index of names. Orientation of a map. Kilometre grid. Principle and use of a compass (magnetism) orientation with a watch |
| 6. Relief | Plains (Great Hungarian Plain), hills (valley, ridge) and mountains. Regions of Hungary. Contour lines and height values | Plains (Great Hungarian Plain), hills (hilltop) and mountains (valley, basin, peak). Regions of Hungary. Contour lines and height values | Plains (Great Hungarian Plain), plateau, hills and mountains. Contour lines and height values |
| 7. Colours on physical maps | Height values *(blue, green, light and dark brown)* | Mountains: brown<br>Hydrography: blue | Legend *(blue, green, light and dark brown)* |
| | Representation of hydrography (spring, brooklet, stream, creek, river, estuary, canal, affluent, etc.) | Representation of hydrography (creek, river, canal, affluent, flow, left and right riverside, lake, marsh, etc.). Depth values | Representation of Hydrography: river, lake, sea, ocean. Depth values |
| 8. Types of maps | Political, physical, tourist, road, city, historical, military, meteorological, phyto- and zoogeographical map. Maps of population density and dialects | Map of provinces, political map, road map, city map, tourist map, thematic maps, atlases | Thematic maps, tourist maps and road maps |
| 9. Map symbols | – | Country and province border, settlement symbol, roads and railroads<br>Mining (coal, lignite, petroleum, natural gas, iron ore, bauxite)<br>Industry (heavy, light, food) | Legend, symbols of routes for excursionists |

(continued)

**Table 15.2** (continued)

| Basics about nature | | | |
|---|---|---|---|
| Title of the chapter/ themes | National publishing house (2002) Our important partner: the map (pp. 104–122) | Mozaik publishing house (2004) Orientation on maps (pp. 34–50) | Apáczai publishing house (2004) Orientation on maps and in nature (pp. 69–84) |
| 10. Map history | – | – | Stick maps, Egyptian map of gold mines, Eratosthenes, Ptolemy, Tabula Peutingeriana, T-O maps, discoveries, Mercator, Lázár secretary, János Zsámboky |
| 11. Others | Satellite, aerial photograph, computer | Satellite image as illustration | Basic rules for excursions in Nature |

source, object-oriented Web application server. A Zope website is usually composed of objects stored in a Zope Object Database (ZODB) and not by files in a traditional file tree system. Zope can be managed using an interactive built-in interface via a Web browser (Plone 2011).

Plone was developed using Python and other programming languages (JavaScript, XML, CSS, etc.). It is multifunctional and can be used for the design of different Web content, e.g. blogs, Web shops, internal sites, etc. Platform-independent (available on Windows, Mac and Linux/Unix) and all its features are customizable, being easy to personalize a website using CSS (Cascading Style Sheets) and add-ons. At same time Plone is one of the fastest open-source CMS platforms on the market.

Plone has also a visual HTML editor and a "drag and drop" reordering and editing tool to create and update content. It supports the creation of backups to avoid data loss. The system has multilingual content management, portlet engine, graphical page editor and so many helpful features. Its use is totally free. These advantages were enough to choose this CMS.

Some noted users of Plone are the Federal Bureau of Investigation (FBI), Brazilian Government, Chicago History Museum, DISCOVER magazine, United Nations, European Environment Agency, NASA Science and Defending Children's rights (Fig. 15.3).

### *15.3.3 Changes in the Structure of the Website*

The structure of the website had to be changed too, to adapt it to the conditions derived from the use of the Plone CSM. The title of each main part was abbreviated

Fig. 15.3 Other websites created with Plone

(shortened) and the menu points were also modified and reduced from eight to six (some of them were joined and others were put under new titles), to fit into the superior section of the website (Fig. 15.4):

- Basic concepts (As you are learning and even more...) (*Alapismeretek*)
- Digital collections (Curiosities and Try out!) (*Digitalis gyűjtemények*)
- Games (Learn and play!) (*Játékok*)
- Events (News and events) (*Rendezvények*)
- Links (What can you find on the Web?) (*Hivatkozások*)—including previous works in Hungarian language
- News (News and events) (*Hírek*)

Asking the pupils about the handling of the website, we could conclude that short texts help to keep awake the children's and young people's interest in the content, re-affirming the principles followed during the development of the first website. However, pupils also remarked that the high number of short pages (almost 100) causes difficulties to access a determined theme within a chapter. Considering this situation, we decided to introduce some changes in the organization of pages

Fig. 15.4 Welcome page of the new website

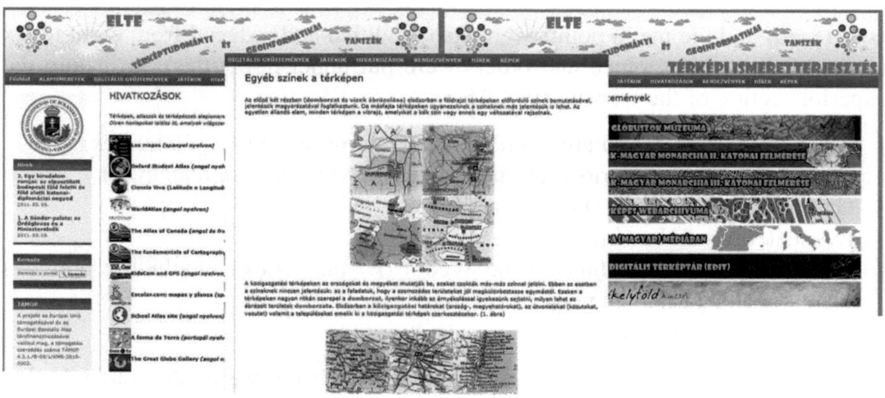


Fig. 15.5 Some pages of the website

within the website, designing longer pages (one by each general theme) and placing more interactive fast links to the main headings (Fig. 15.5).

Other important change made was the addition of a search tool, which is constantly accessible from the top of the website, to seek and locate information within the homepage. This is a useful tool asked very often by the user children in our meetings during the last years. Other, a most visual innovation was the graphic and dynamic (real-time) presentation of the number of visitors to the website, using an Earth globe and/or map designed and maintained across the www.revolvermaps.com website.

The online-glossary created in 2000 was also changed, adapting it to the new environment, but keeping the same content and assuring the fast access to the information when the user moves with the mouse over an unknown concept. The use of a glossary fills two aims:

- To explain concepts related to other sciences, concepts that children could not learn between third and sixth grade of Elementary Schools (e.g. magnetite, used in the explanations related to the compass)
- To explain concepts presented in other themes, avoiding the links between different pages and themes, to keep the continuity of reading the actual page. E.g. Equator, a concept mentioned in the first theme ("Map and Reality"), but explained in more detail in the ninth theme ("Latitude and Longitude").

## 15.4 Future Plans

### *15.4.1 Testing the Grade of Acceptation*

Participants in the present project are interested to know the critics, suggestions, etc. from pupils and teachers. The more directly way of contacting us is across the website, sending their messages to our email addresses. We are also considering the option of creating at least two Web based questionnaires, one for pupils and other for teachers. These questionnaires will ask about the design (emphasizing this aspect in the questionnaire for children) and content (emphasizing this aspect in the questionnaire for teachers) of the website.

### *15.4.2 Versions in Other Languages*

The Plone CSM includes a choice to make an automatic translation of the pages into other supported languages. The translation can be made using LinguaPlone, a tool that in March of 2009 supported more than 40 languages apart from English (French, German, Spanish, Italian, Portuguese and others).

The main difficulty of translating the original Hungarian text into other language is not the translation itself, but the correct interpretation of the text related to cartographic concepts. Other difficulty can be that children from various countries speaking the same language can use the same words with a different meaning or can name the same object using different substantives that have different meanings in each country. Specialists should also consider that a considerable percentage of the illustrations containing maps have to be substituted, because these figures were designed for Hungarian pupils after being selected from Hungarian School Atlases.

### *15.4.3 Connecting the Website to Social Networking Websites*

Social networking websites play a very significant role in the more effective communication with the younger generations, which are growing living in a virtual community grouped within websites such as Facebook, Tagged, hi5, Hungarian Iwiw, etc. and websites offering social networking and microblogging services, like Twitter. Considering our interest to keep Hungarian pupils informed about the newest updates to the website, we are considering seriously to create profiles in Facebook and Twitter as "indispensable" tools to maintain young people informed.

### *15.4.4 Other Ideas*

The structure of the website created with Plone CSM lets us widen the content with new themes if needed and to keep the information updated in the server of the Department of Cartography and Geoinformatics (Eötvös Loránd University). At next future, we plan to add more examples developed at the Department and more interactivity to the website, taking advantage of the choices offered by CSM as well as considering the suggestions to be made by Hungarian pupils and teachers in the interest of maintaining a website that increasingly meets their learning and teaching needs.

**Acknowledgements** The project is supported by the European Union and co-financed by the European Social Fund (grant agreement no. TAMOP 4.2.1/B-09/1/KMR-2010-0003).

## References

Cartographia TK (2008a) Első Atlaszom. Cartographia Tankönyvkiadó Kft, Budapest

Cartographia TK (2008b) Földrajzi Atlasz a 10–16 éves tanulók számára. Cartographia Tankönyvkiadó Kft, Budapest

Cartographia TK (2008c) Középiskolai Földrajzi Atlasz. Cartographia Tankönyvkiadó Kft, Budapest

Hartdégenné RÉ, Köves J, Rugli I (2010) Természetismeret az általános iskolák 5 évfolyama számára. Nemzeti Tankönyvkiadó, Budapest

Horváth M, Molnár L, Szentirmainé BM (2002a) Természetismeret az 5. osztály és a 11 éves korosztály számára. Apáczai Publishing House, Celldömölk

Horváth M, Molnár L, Szentirmainé BM (2002b) Természetismeret munkafüzet az 5. osztály és a 11 éves korosztály számára. Apáczai Publishing House, Celldömölk

Plone CSM (2011) Plone CSM. http://plone.org/. Accessed 18 April 2011

Reyes JJ (2002a) Térképészeti alapismeretek bemutatása a weben. Dissertation. Eötvös Loránd University, Budapest

Reyes JJ (2002b) Iskolások számára készített honlap a térképekről. Földrajzi Közlemények, Vol. CXXVI.(L.), number 1–4:130–138, Budapest

Reyes JJ (2006a) A website about map history for Hungarian pupils. In: Proceedings of digital approaches to cartographic heritage, 77–82, Thessaloniki

Reyes JJ (2006b) GIS site for Hungarian pupils. In: Digital proceedings of the first international conference on cartography & GIS, Borovets, Bulgaria

Reyes JJ (2006c) Multimedia cartography for Hungarian pupils. In: Proceedings of modern technologies, education and professional practice in geodesy and related fields, 240–248, Sofia

Tompáné BM (2002) Környezetismeret 4 (Élő és élettelen környezetem. Élet a ház körül. Környezetismeret tankönyv 10 éves tanulók részére). Pauz-Westermann Publishing House, Celldömölk

# Chapter 16
# Teaching Cartography to Children Through Interactive Media

**Marli Cigagna Wiefels and Jonas da Costa Sampaio**

**Abstract** Computer technology arrives very quickly to schools where young students from the first grades already utilize it for electronic games, for access to internet or for video games, among friends or in family. The proposal to build an interactive media with the basic information about the city of Niterói (State of Rio de Janeiro) was the first step of an activity intended to help geography teachers in their daily work. The municipality of Niterói, located in the State of Rio de Janeiro, has 458.465 inhabitants (2001) and is the scenario of all geographic information provided by this media. The location of UFF in this municipality and the edition of the school Atlas of the municipality of Niterói, implemented during a former university project, constituted the main inspiration for the present project. Students and school children were mobilized for the project and contributed providing information for the media. All begins with a mental map of the schoolchild going to school, with different graphic situations: the street of the school, the school itself, the classroom. It is in these places that all events of the media happen. According to the Brazilian National Programme parameters (Brazilian Ministry of Education (2011) Parâmetros Curriculares Nacional (PCN).http://portal.mec.gov.br/seb/arquivos/pdf/livro01.pdf and http://portal.mec.gov.br/ seb/arquivos/pdf/geografia.pdf. Accessed 20 June 2011) one should "know how to use different information sources and technological resources in order to build the elements of knowledge". This is very propitious to the introduction of this media in the first grades of the basic school. The building of mental maps, the understanding of photographic images, orientation, environment and the graphical recognition of objects are the main themes treated in the media. In the present project, school and graphic cartography provide the young pupils with a material which stimulates them to learn and with recreational activities which strengthen their liking for geography and cartography. At the end of the project, the media will be open to

M.C. Wiefels (✉) • J. da Costa Sampaio
Universidade Federal Fluminense, Rio de Janeiro, Brazil
e-mail: cigagna@vm.uff.br; jonasjjj@yahoo.com.br

L. Zentai and J.R. Nunez (eds.), *Maps for the Future*,
Lecture Notes in Geoinformation and Cartography 5,
DOI 10.1007/978-3-642-19522-8_16, 

public on the university web site and dissemination seminars will be provided to school teachers of the municipality.

## 16.1 Methodology

### *16.1.1 The Idea*

"The classroom in a shoebox": The idea appeared during the preparation of a model of a classroom which was materialized in a shoebox. The model allowed the visualization of the physical elements of a classroom according to a vertical spatial projection.

The elements of a classroom in the first grades of the basic school are very propitious to the project: they constitute significant fixed elements like the student desks, the teacher desk, the blackboard, wall charts, the bookshelf, door and windows. On the teacher desk we find a terrestrial globe, books, pieces of chalk, a cleaner. On the sideboard in the corner, we find a small museum nearby an aquarium. Based on these elements, other icons will be presented and activated in the media.

The activity will be developed based on the classroom. Before to arrive, in the classroom other scales are presented, as the building of a mental city map, the arrival at school and finally the classroom.

For this purpose, many proposals were suggested and selected (Nogueira 2009). A bibliography specialized in school cartography was consulted and surveys of electronic games were necessary to perform the different stages of the media (Pontuschka et al. 2007).

The presentation of the activity begins with a sketch drawn by a child (mental map) symbolizing the way from home to school. In a second moment we see a bus passing in the school street (Fig. 16.1). It stops and a child gets out. He goes until the pedestrian crossing and waits for the green light to cross. He is now just at the entry of the school. During these stages, we see several situations that help in acquiring knowledge. The teacher will be able to treat several aspects of localization, orientation, and space with his pupils. He will also deal with the location of the school in the district, the orientation of space according to the pupil's home and the representation of the near space related to the far space. All different special points nearby the school will be observed and lived in the media.

### *16.1.2 Organization and List of the Themes*

The themes selected for the media are: the Earth references, the rotation and translation of Earth, the orientation with the compass, the cartographic and historical representations of Niterói, the landscape, the coastal fauna and environment.

**Fig. 16.1** The bus run to go to school

The presentation of these themes proceed from the fixed elements in the classroom, as for instance the terrestrial globe on the teacher desk, which activate many windows informing about geographic concepts and suggesting games to pupils with demonstrative icons.

### *16.1.3 Human Resources*

Building up the team is the most important stage as well as the most difficult, mainly when it is about preparing activities for children. The project is performed as a University extension activity, entitled "Production and dissemination of didactic material for the teaching and practice of Geography". In the team, we have a scholarship student from the computer science course and three scholarship students from the secondary school programme "Young Talents for Science".

> The Young Talents project is a scientific pre-initiation programme implemented since 1999. This project is realized in partnership with research institutions and with public and private universities of the State of Rio de Janeiro. Through this mean, students of the public secondary and professional schools are inserted in scientific pre-initiation. (CECIERJ Foundation)

The "young talents" Felipe Correa Mesquita, Kenya de Oliveira Silva Souza and Maicon Peclat are students of different secondary schools of a Niterói neighboring municipality and participate for the second consecutive year in the researches for this project. They collaborated with the authors in graphical work and in the design of the media, as well as in bibliographic research.

### 16.1.4 Resources and Institutional Support

The project is implemented at the UFF's Institute of Geosciences, in its cartographic laboratory, with the direct support of the Department of Geo-environmental analysis and of the UFF's Board of Extension, where the project is registered.

The project is very ambitious and the financial resources are limited to the scholarships awarded to the students. It has a great educational value as it makes secondary students work together with university students. As a matter of fact the project has begun through an exchange of ideas between the members of the team. The data inserted in the project were provided by the students of the project team during their researches and of course also by the authors.

### 16.1.5 Creating and Realizing Activities

This includes the design of the boards as well as the illustrations included in the work, the supporting pictures, and the researches about the contents, the conception, the graphic programming, the animation, and the final arrangement.

### 16.1.6 Evaluation

The evaluation of results will be done in schools where the media will be tested. The results will be presented in the academic agenda of the UFF during the Science and Technology week of the University. The results will also (and mainly) be presented to the teachers of the public schools network during a special seminar.

## 16.2 Development of the Media

The planning and execution of the project are based on the content of the media with a special language adapted to the children of the first years of basic schools. A special attention is given to the content and to the way this content can interact in the education and in the learning motivation of geography and cartography. The progress attained in this training leads the children to be more reflexive and clever in the content games and in taking notes. The selection of themes is strongly associated to the "fix points" located in the classroom space.

The objective of the media is to allow a constant dialogue with the "playing" child, who will be able to play alone at home or in the school computer room with the presence of a teacher who can intervene and orient him in the utilization of this pedagogic tool.

Fig. 16.2 Game: crossword puzzle

By clicking or by passing the cursor on the image, many elements are activated and a window opens in order to inform about the next steps of the proposed activity. At this moment information and concepts are proposed followed by games. The time taken to perform the games is measured and helps the player to evaluate himself (Fig. 16.2).

## 16.2.1 The Creation of Themes

Themes are grouped according to significant elements in the classroom (board, teacher table, pupils' desks...) that we call "fix points". This group of objects allows the building of the contents.

The different points of view of the classroom image were graphically designed based on pictures taken in a model classroom located in the premises of the UFF's Institute of Geosciences. These pictures were registered in different angles. This allows the vision of space and environment according to different perspectives.

## 16.2.2 The Earth Rotates (Rotation and Translation Movements: Seasons of the Year)

These themes are developed based on the board "birthdays this month" located near to the door. The media, in its basic conception stimulates the ordinary daily life of a classroom in the first years of basic school.

In order to highlight the theme "The Earth rotates" we begin with a Globe rotating from west to east, simulating the alternation day/night. Information about this movement is presented since the beginning of activity. Passing the cursor on the birthday board presents the months of the year as well as information about the seasons. This part of the media is more informative in order to satisfy curiosity. It presents the interest of the media dynamism and allows a supporting interaction of the teacher.

On the same wall of the classroom we find a bookshelf with books and maps. These elements are enabled to execute the media activities. The game box for instance can be enabled and the activity hereunder will appear on the screen.

Continuing going around the classroom, the pupil will be stimulated top observe the terrestrial Globe on the teacher's desk and, by passing the cursor on it, to activate it. New windows will then open with the presentation of different representation concepts of the Earth, by means of terrestrial globes and maps. The reference lines of Earth are indicated as well as the presentation of meridians, parallels, latitudes and longitudes.

After the presentation of cartographic concepts and the illustration of the main reference lines of the terrestrial globe, the pupil will be invited to participate in a game of identification of geographic coordinates on a map where small ships appear. The pupil has to indicate their coordinates (Fig. 16.3).

All games count points and have a limited time to be realized. This is already commonly what happens with electronic games. The timing in the different steps

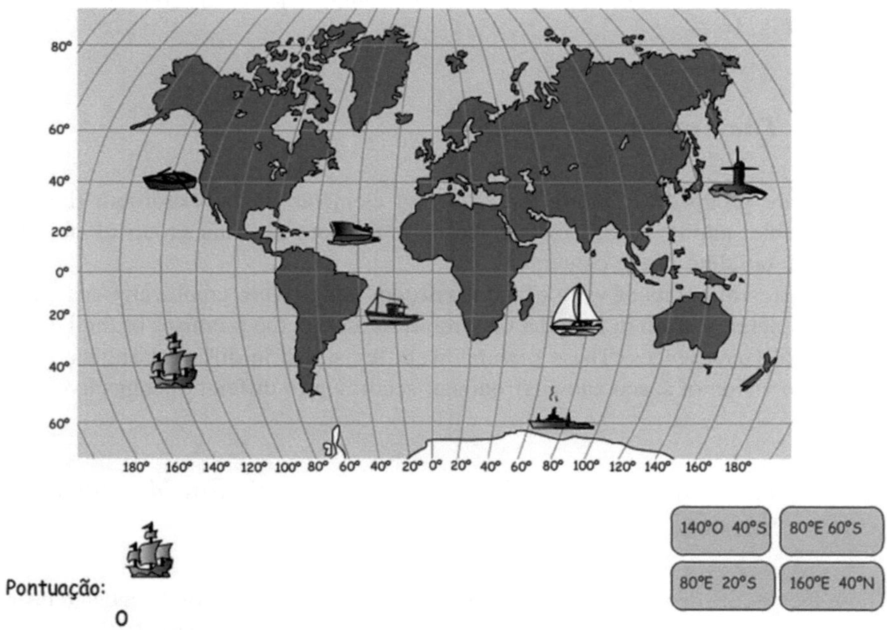


**Fig. 16.3** Localization game: where are the ships?

allows competition among pupils and encourages their rapidity and smartness. That is why the project team thought about introducing a chronometer.

The pupil will always be invited to develop his knowledge and to participate to the several activities proposed by the media during his virtual tour through the classroom. In this school environment, designed for learning geography, we use an attractive visual and graphic language, in order to provide a more friendly and interesting training, and to offer challenges to advance in the reading of maps and in practical cartography.

Other important fixed points in the media are the classroom windows. It is through these windows that the landscape can be observed. The landscape can be the community, the other side of the school building, the urban landscape or even a beach. Regarding this theme, we try to stimulate the sketching of the landscape presented by pictures of Niterói, cut in nine equal rectangles, making it possible to build a puzzle as a linked activity.

Another fix point is the waste bin in the corner of the room, nearby the teacher desk. Pupils are invited to read short texts about garbage, their decomposing time and the care to be brought to environment. Afterwards they are invited to a selective collection of garbage, with bins of different colors (green, yellow, blue and red) according to the type of garbage to throw.

In the classroom we also have a small museum with relics of a thematic long walk along seaside, and an aquarium with a fish. The question put to the pupil: is the fish is its natural environment? At this moment an activity is introduced with information about the fauna (fishes, crustaceans, molluscs and marine mammals) of the Guanabara Bay. The proposed game afterward is very like a "naval battle", using alphanumerical coordinates to localize each element of the game. At each good shot, the player receives a picture of the animal (fish, crustacean, mollusc). In this game the animals are only presented with pictures.

The daily classroom goes this way and the cards and posters stuck on the board are elements of activities for these media players. The pupils will have the opportunity to make their own board posters according to the suggested themes.

## 16.3 Conclusions and Perspectives

> Games constitute a resource still seldom used in classrooms. They are however very valuable because they arouse expectation, anxiety and enthusiasm to the pupils. A game is ludic by itself and presents challenges well accepted at all ages, in the classroom as well as outside it. For students, it is something amazing, as the game sounds like a challenge to their cleverness, bringing them to know more about the rules and to think about winning strategies. (Passini et al. 2007)

The didactic resource represented by this media and described in the present document is the result of a research and of a dream. We have brought together a heterogeneous group of students to build an educational and interactive media, where concepts and actions are mixed in order to allow the pupil training through interesting activities and games.

The new communication technologies are essential instruments in a classroom where teachers and pupils are integrated and make instruction and education a dynamic and friendly training. Who does not like to have a game in a classroom? However, we must also have in mind that not all teachers already live in a digital era and that not every school has its own computer room. Our proposal is therefore to have this media available as an important auxiliary tool in a classroom for the teaching of the proposed themes.

Once we have completed this first model, it will be evaluated by the teachers of the public schools network, as well as by those of the private schools, in order to improve it even more.

## References

Nogueira RE (2009) Motivações Hodiernas para o Ensinar Geografia—representações do espaço para visuais invisuais. Florianópolis

Passini EY, Passini R, Malyz S (2007) Prática do ensino de geografia e estágio supervisionado. Contexto, São Paulo

Pontuschka NN, Paganelli TI, Cacete NH (2007) Para ensinar e aprender Geografia. Cortez, São Paulo

# Chapter 17
# Cartographic Response to Changes in Teaching Geography and History

**Temenoujka Bandrova**

**Abstract** We live in a time when many GI specialists recommend the usage of electronic maps and atlases and will not develop traditional cartography based on paper version products. This report considers both aspects: why we are not ready to use only electronic versions of maps and atlases, and how we can improve the traditional cartography by introducing new topics and modern visualization. Several examples are given from the latest school atlases that are used in the education of geography and history in Bulgaria. Some new aspects in cartographic products and visualization were created because of the permanent changes in school curriculum and in the content of the geography and history atlases. Experiences and research in schools motivate cartographers try to improve the maps and atlases.

## 17.1 Introduction

The specialists working in cartography should respond to the everyday changes of life. However, several months or years are necessary to enter changes in the school curriculums. New important topics appear so quickly that the cartographic products illustrating them become outdated by the time they come out of press. Were we ready to respond to Haitian children after the earthquake that struck the country on Jan. 10, 2010? "*A study by the Inter-American Development Bank estimated that the total cost of the disaster was between $8 billion and $14 billion, based on a death toll from 200,000 to 250,000. That number was revised in 2011 by Haiti's government to 316,000*" (The New York Times 2011). Are we ready to respond to the technically most developed nation and its children—Japanese children? Let us see what happened in the first 3 months of 2011: floods, earthquakes, landslides! Some

T. Bandrova (✉)
Department of Photogrammetry and Cartography, University of Architecture, Civil Engineering and Geodesy, Sofia, Bulgaria
e-mail: tbandrova@abv.bg

L. Zentai and J.R. Nunez (eds.), *Maps for the Future*,
Lecture Notes in Geoinformation and Cartography 5,
DOI 10.1007/978-3-642-19522-8_17, © Springer-Verlag Berlin Heidelberg 2012

**Table 17.1** Natural disasters in 2011 (January–March), data by McKibben (2011)

| Types of disaster | Place | Time, 2011 | Human cost | Economic cost |
|---|---|---|---|---|
| Earthquake of magnitude 9.0, followed by a 15–20 m high tsunami | Japan, north-east coast | March | More than 10,000 dead; 17,000 missing | € 215 billion |
| Landslide | Brasilia | January | 916 dead; 345 missing | € 213 million |
| Floods | Australia | Nov., 2010–January | 37 dead, nine missing | € 22 billion—Australia's costliest natural disaster ever |
| Earthquake of magnitude 6.3 hit the city of Christchurch | New Zealand | February | 166 dead | € 5–7.6 billion |
| Devastating floods | Sri Lanka | January–February | 62 dead; 1.1 million displaced | € 340 million |
| Earthquake of magnitude 6.8 | Myanmar (Burma), 30 miles north of Tachileik on the Thai–Burma border | March | At least 75 dead; more than 110 injured | Not known in April 3rd, 2011 |
| Floods, heavy rains continued from December last year | Philippines | January–March | At least 75 dead | € 30 million |
| Severe storms, lightning and floods | South Africa | January | 91 dead, 321 injured | € 83 million |

scientists and journalist already defined 2011 as "a year of disasters". McKibben (2011) gives us some data, which are summed up in several countries and millions of people were affected during the first 3 months of 2011 (see Table 17.1). How cartographers will respond to children from Japan, Brasilia, Australia, New Zealand, Sri Lanka, Burma, Philippines, and South Africa? In these cases, it is clear that we are not ready to give response, we are too slow to change something. We are not ready with standards and visualization of those geographical process that are happening around nowadays. These statements make it evident that more scientists are needed to work on these topics, more projects should be financed and developed, more governments should help the scientists to develop standards and fast responses to everyday changes. They should give enough information and knowledge to our children and pupils and prepare them for the right behaviour.

Natural disasters are only one of the many topics that should be dealt with in the geography and history subjects of school curriculums (e.g. globalization, sustainable development, environmental protection). They are relatively new, and they find their place in the changing school curriculums. Cartographers try to respond by including such themes in their new cartographical products, which are used as illustrations in the geography and history lessons. The problem is that these cartographical products need to be updated more often than until now.

## 17.2 Changes in Geography and History Contents

The Ministry of Education in Bulgaria is responsible for the creation and approval of school curriculum for all subjects in Elementary, Secondary and Higher Education. Here I will describe the situation in Geography and History subjects. The changes are happening every 5 to 10 years. This situation is not typical of Bulgaria only. (Zhang and Foskett 2003) declare that the changes are in the subject matter in 15 sets of British geographical textbooks in the UK from 1907 to 1993. Some of the changes are caused by the shift from general to regional aspects, just like the concept of "regions" in elementary and secondary schools in the USA (Stoltman 1992). These changes led to new aspects, details and topics in teaching. Every new curriculum is made to improve education and give children new aspects of a modern society. In many countries, this is done to provoke independent and active learning (Paris and Byrne 1989).

A new law about Elementary, Secondary and Higher Education in Bulgaria has been in preparation. This will bring changes in the geographical and historical curriculums. Many changes were introduced also in last few decades. The biggest ones were in the general context of the mentioned subjects. The school subject of Geography changed its name to "Geography and economics", and History to "History and civilizations". These new scopes developed many topics that enlarged 'Geography' and 'History' to become broader subjects including the sustainable development of the society. The content in geography was enriched with some new topics like risk management, environmental protection and ecology. New topics have also arisen in History. For instance, we could present a visualization of the Trojan War and the brigade distributions in the socialistic time.

## 17.3 Cartographical Response to the New Aspects in Geography and History Subjects

To help geographical and historical education, cartographers propose various types of cartographical products: wall maps, atlases, electronic atlases, contour blind maps, globes, virtual cartographic representations, etc.

### 17.3.1 Kindergarten

The first questions could be "when should we start?" There was a discussion between the old cartographic generation keeping positions in the state cartographic company and the new generations working for private companies. Before 1990, the pupils faced the first map in the geography lessons in the third grade (10-year-old children). Nowadays, children receive everyday information by the media, radio, TV, Internet and they start to use and read maps in their kindergarten age. The fact that young kids are able to use maps was proved by experiences with children in 15 kindergartens in Sofia. The children had knowledge about the names of continents, Bulgarian boundaries and cities, the names of the biggest mountains and rivers. They were very active and accepted the questionnaire as a pleasant game. Based on this experiment, we created two posters for kindergarten:

1. Plants and animals in the World, and Plants and animals in Bulgaria on the back side of the poster (Fig. 17.1).
2. The major sights in Bulgaria, and Six great Bulgarian personalities of our history on the back side of the poster (Fig. 17.2).

Our work for and with the children in kindergartens aims to achieve results in several aspects:

- Prepare children for their first school lessons which need the use of maps;
- Improve their geography culture;

**Fig. 17.1** Plants and animals in the world

**Fig. 17.2** Major sights in Bulgaria

- Give them additional knowledge in a distracting way, namely, by playing games and drawing pictures;
- Show them the artistic aspect of the maps and train them in the aesthetical view of maps.

All participants in this initiative believe that these kids will learn how to use maps in an appropriate way. Facing the maps prepared of high quality, teachers and kids will be able to distinguish the good and useful products from the low quality ones, which are sold on the free market.

### *17.3.2 Electronic Versions of Cartographic Products*

Pupils and kids share the values of a new generation, which is using computers every day for every need. This is why the most appropriate solution is to propose geographical and historical mapping visualization through the medium of electronic cartographic products. Several ideas have been realized worldwide. An electronic school atlas of Quebec is a good example as a product: its structure and content is related to the child (Anderson et al. 2003). Other research (Pfander et al. 2004) represents the Arizona Electronic Atlas (http://atlas.library.arizona.edu) as an interactive Web based state atlas, which allows the users to create, print or download maps and data. The creators focused their efforts on the integration of the Atlas into the university classroom.

There are different opportunities for pupils to get trained in organizing information according to the classification schemes, understanding intellectual property rights and mastering the use of Internet based tools in the Norwegian GeoAtlas (http://www.avinet.no/). Many examples of electronic atlases are available on the market and in the web. Nevertheless, only a few of them are developed for use by the children and pupils. There is a large number of advantages of these atlases. They are well described by Uluğtekin and Bildirici (1997). One of the biggest advantages is shown by the same authors as they wrote, “Atlas publishing in electronic form will become financially interesting more and more”.

After this introduction to the electronic atlases development in the world, let us pay attention to the Bulgarian version of such an atlas called “MaxInfo” produced by Datamap-Europe Ltd. It contains a lot of information about Bulgaria, Sofia and other Bulgarian cities: maps, legend, photos and texts (Fig. 17.3). The information can be visualized on computer screen and can be printed. Teachers and pupils can include the necessary objects by interactively chosen symbols and automatically place them on the base map. The pupils can search the necessary information by typing a keyword and locate or see it in pictorial and/or text form. In this way, they get knowledge about the GIS function in the first step of their education. The Atlas

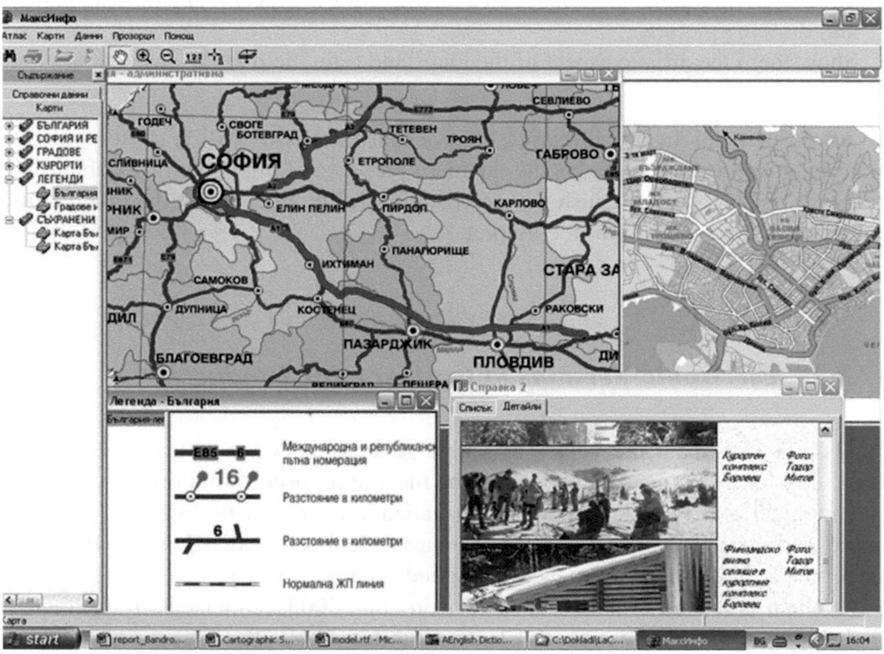


**Fig. 17.3** A possible use of an electronic atlas is combining data

is working in a user-friendly way, so the pupils will begin to use GIS questions and will understand the productivity and usefulness of the real GIS (Bandrova 2001).

The training with electronic atlas could be directed in three ways:

- Work with data. Different kind of tasks can be given to pupils. They can select a list of type objects, classify objects, examine some details of every object, and connect them with information for other objects. It is a flexible system for object searching according to different criteria—name, type, address, key words and others. Every reference can be saved for future usage.
- Work with maps. It is possible to scale maps, move map images, switch on or off the visibility of layers, and choose a symbol system for object mapping.
- Work with data and maps. Pupils can add data on a map as a separate map layer, receive information for objects, find the object situation from a data list, etc. Examples of the teachers' tasks can be the following according to Ormeling (1996):

*1. What is on a map (identifying);*
*2. What is where on the map (classifying);*
*3. Do you see a relationship on the map (relating);*
*4. Check if this relationship is valid for each region on the map (checking, monitoring, validating).*

This electronic atlas has a CD version and it is sold on the Bulgarian market. This makes it difficult to be used by the teachers and pupils. The main users come from the business field. The pupils use the free version of another Bulgarian web based atlas. It can be found on www.bgmaps.com, but it is not adopted for pupils or children use. Its main purpose is to give the possibility of finding addresses in a city.

### 17.3.3 Compromises?

Compiling and mapping Geography and History information, which we needed for school atlases and wall maps, provided us with enough data that allow us to produce electronic or virtual variants of these cartographic products. From the cartographer's point of view, this will not be a problem. Also in these days many variants of electronic textbooks were published. It is clear that this will be the future of school cartography. However, the situation is still bad because of the poor computer equipment in the classrooms. Only one room per school is equipped with computers. This room is all time occupied by pupils studying computer science. There are no chances for geography teachers to use this room for their lessons. For this reason, we still propose producing only paper atlases and wall maps for Bulgarian market. However, in outline maps and in some atlases there are tasks that require the users to perform Internet research. Another reason is the open issues in terms of legacy. The pupils will use their personal notebooks, but we cannot allow the usage of illegal software. Unfortunately, the necessary software for cartographic applications is too expensive. This point of view has been developed in the work of Gold (2008). He states that:

> However, where vandalism, or irresponsible use such as software piracy, are perceived to be problems with the result that the machines may only be used when staff supervision is available, then even a large number of microcomputers in a laboratory may provide extremely limited access to students taking computer cartography courses.

A good idea is that Geography and History lessons could be developed as computer games appropriate for different ages. Such experiment could be found in The Serious Games Institute (SGI) at Coventry University. Wortley (2008) says that it is established to "*create an international centre of excellence for serious games and virtual worlds. The SGI has been pioneering the use of virtual environments for a range of applications which include e-learning, simulation, disaster management, virtual conferencing and social and business networking*". Here, the issue is how to make such products cheap enough or free of charge for teachers and pupils and how to make them attractive for the pupils' needs.

### *17.3.4 Modern Visualization*

The visualization of geographic and historic information in the school atlases is one of the main topics of modern cartography. Mapmakers should reach the pupils' attention and provoke map reading and analysis. How do the children think in terms of cartographic aspects? We can find different responses to this question as well as many proposals. Many of these proposals are the result of questionnaire research. Based on international research and questionnaire, Reyes Nunez et al. (2005, 2008) propose the use of "*satellite images in the textbooks and atlases. . ., which help the pupils to understand the content of the physical maps by visualizing the represented territories in their natural dimensions*".

How to visualize the necessary information? How to choose the appropriate symbol system? What colours should we use? The answers to these questions need experiments with children to understand their perception of maps (Konecny and Svancara 1996). In the work of Bandrova (2007), one of the main steps of the proposed technology for atlas creation is to work with pupils and to be close to their way of thinking. After many experiments conducted mainly by questionnaire, the prepared maps and atlases were approved and later published (Bandrova 2010).

There are several major differences between this atlas and the old atlases available on the Bulgarian market:

- Usage of many photos and drawings in the map and atlas design (Fig. 17.4). This attracts pupils and direct their attention towards the topic of the maps.
- Use of various diagrams, statistics and texts like in an encyclopedia style (Fig. 17.5). These not only textual but also visual elements capture the pupils' attention, provide them with more specific and curious information that is easy to understand and memorize.

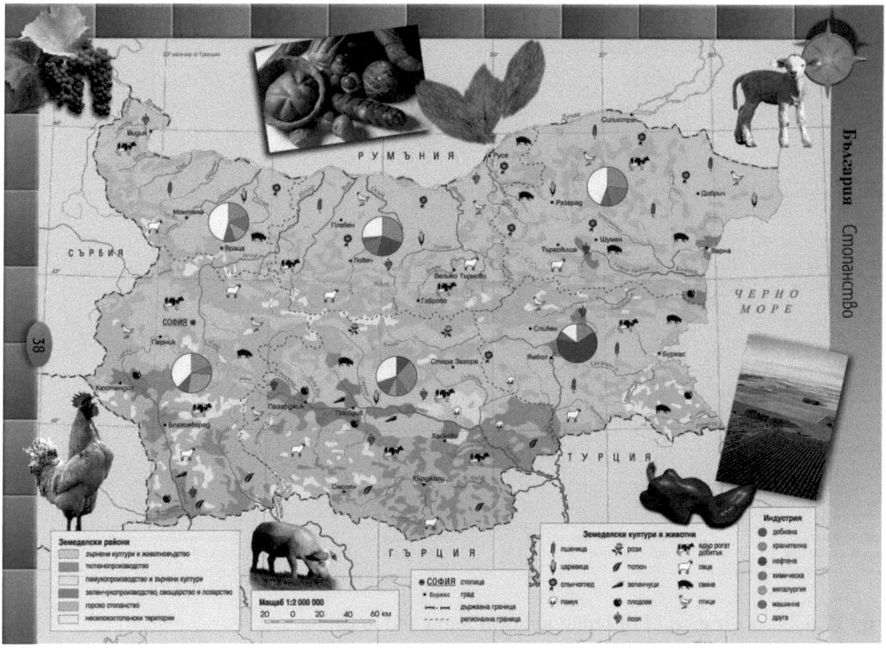


Fig. 17.4 Use of photos and drawings in the map design

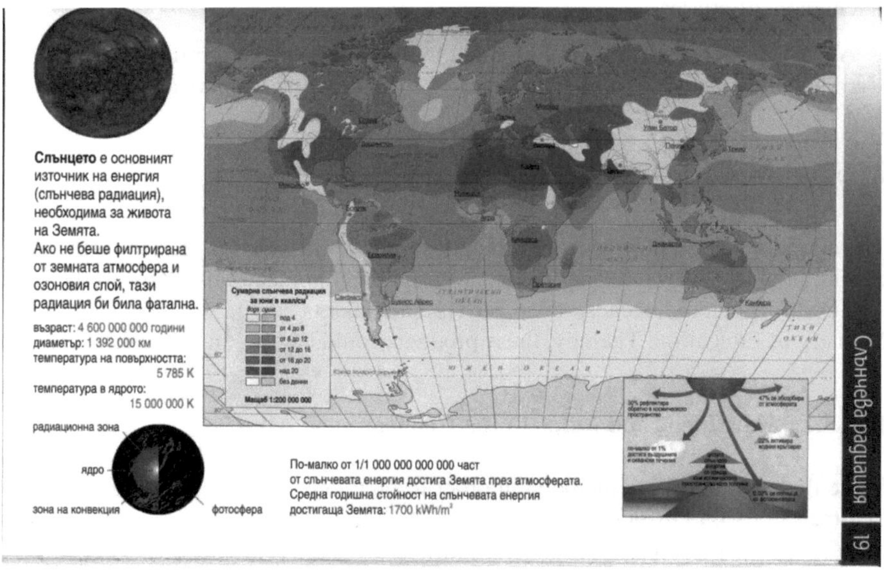


Fig. 17.5 Use of diagrams and text in the map design

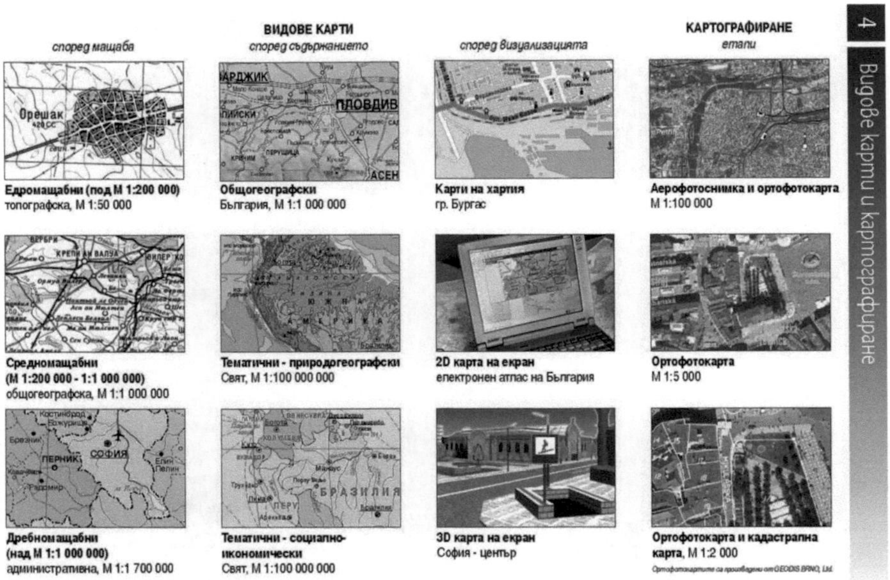


**Fig. 17.6** Visualization of different kinds of maps and mapping

- Use of examples in topic explanations. This method was used to explain different types of mapping and maps (Fig. 17.6).
- Use of colours on the borders of the pages or under the title to indicate the chapters of atlases.

All these improvements accompanied by the high quality printing and publishing make the atlases popular and attractive for the pupils, parents and teachers.

## *17.3.5 New Topics*

The adoption of the new topics in the atlases and wall maps is related to the new curriculums in geography and history. Cartographers must adjust their products to the new curriculums because their cartographic assets should be approved by the Ministry of Education. Thereby, these products could be used in Bulgarian schools.

### 17.3.5.1 New Ideas and Maps in the Geographical Atlases

Some the new topics are included in order to facilitate the teaching of other topics that the pupils find difficult to understand, such as different kinds of mapping and maps (Fig. 17.6) or map projections (Fig. 17.7).

Another reason is to help teachers and pupils to understand themes that are more complicated in their lessons. Map projections represent such a challenge. We try to

**Fig. 17.7** Map puzzle is used in lessons on map projection and deformations

show the pupils how we manage to display the sphere on a plane. Different kinds of deformations appear (area, linear or angular) because of this transformation. This is visible in the map game included as a separate part of the Atlas on Geography—a puzzle or an icosahedron (volumetric map of the world with 20 equal triangular parts in gnomonic projections). Playing with them is multifunctional in terms of geographical, cartographical and mathematical aspects (Fig. 17.7).

Other new topics that appear in the geographical atlases are related to natural risks and disasters, natural resources, ecology and others.

Such examples could be seen in Fig. 17.8. Children and teachers are prompted to describe and analyze the most affected territories and spots where disasters are likely to happen or some consequences that have already appeared. On the basis of such maps some preventive activities could be planned and discussed.

The cartographer's role is to help children, pupils and teachers in their educational process. Another goal is to direct the pupils' attention to the global and regional problems and to the ways of improving and saving the environment. The next task is to find data and represent them in regional atlases at larger scales so as to help regional geography items too.

#### 17.3.5.2 New Ideas and Maps in the Historical Atlases

The old historical atlases were criticised by teachers mainly in the following directions: too many represented objects from general geographic base, a lot of information represented in atlases for young pupils (e.g. the history maps of Bulgaria for school years 5 and 11 have similar contents), too many symbols in the legend. The authors tried to avoid these difficulties, and the new historical atlases are compiled in a different way:

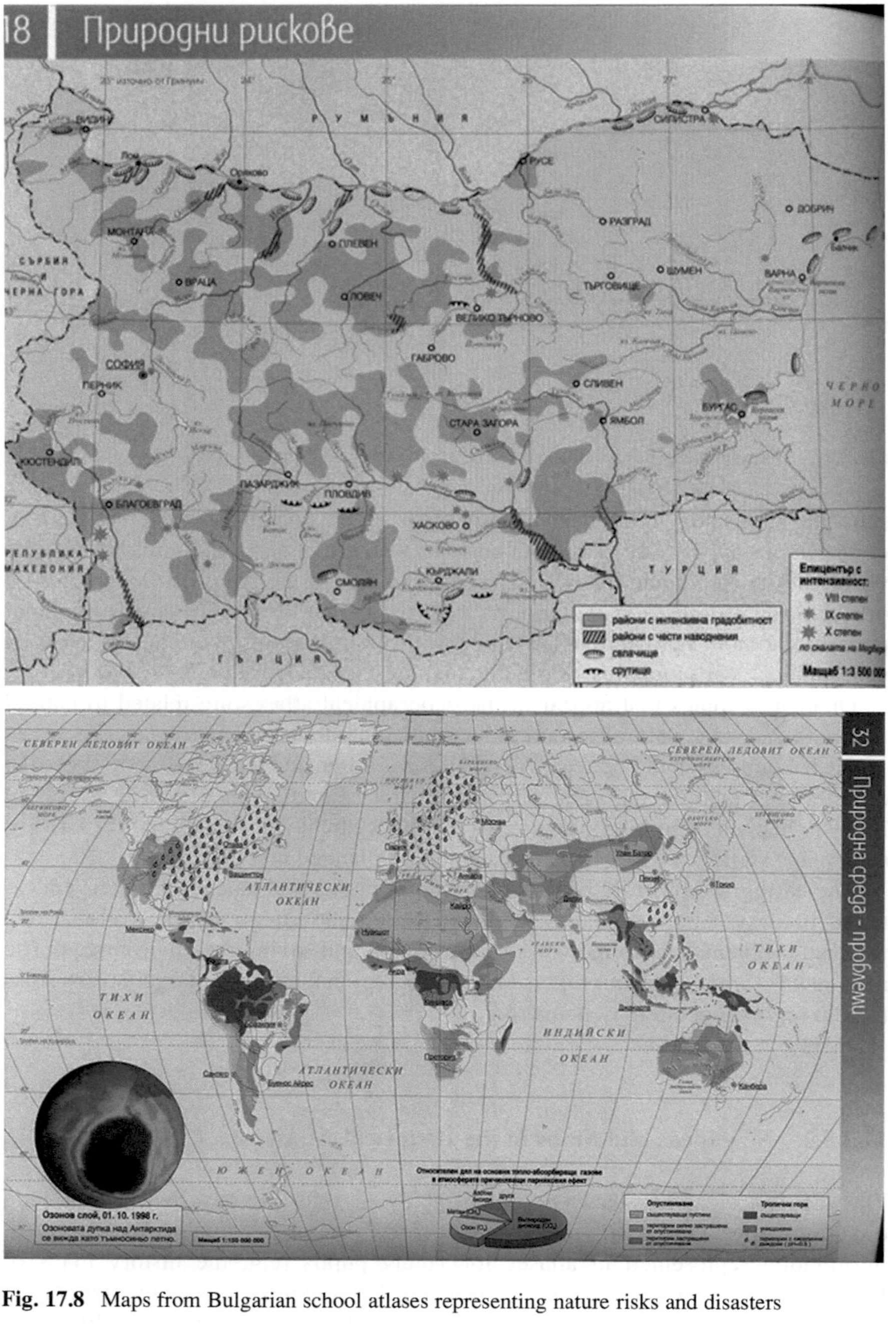


Fig. 17.8 Maps from Bulgarian school atlases representing nature risks and disasters

- Clear general geographical base: only the biggest rivers and those relevant to the topic are represented, and relief shading has replaced a lot of mountain names;
- Different representation of one and the same theme for the different school years: maps and atlases compiled for higher classes have richer content of the topics;
- Not so many symbols are represented in the general legend (only the most associative ones, such as the symbols representing state border, city, capital, etc.).

The new philosophy of atlas compiling is related to the representation of specific themes in the Atlases. Such a theme is the Trojan War, which is not represented in the old atlases. The maps is based on the information taken from Homer's poem "Iliad". The result is shown by Fig. 17.9.

The existing maps representing the Trojan War did not provide us with the necessary information. The war was only partly represented or with insufficient information. The explanations in textbooks should be expanded with descriptions taken from literature. The teachers are very positively surprised when they receive such maps that they use to visualize their explanations.

Other new topics in the historical atlases are related to modern history. Brigade movement is one of the topics that are newly developed. These new maps re-group the distribution, and the localization of different meetings, grouping activities, protests and others are represented. The pupils visualize the topics and they can use the maps to connect history lessons to situational facts.

## 17.4 Conclusions and Directions for Future Work

The idea of this research was to find topics in the school curriculum of geography and history, where cartographers should quickly react. The best solution is when the people in the responsible ministries react also in time and do not take a year or more. However, if it is not possible, cartographers could propose some new themes as additional topics. This is allowed because about 20% of the content of school atlases should contain information that are not included in the school textbooks. The most important topics in these days are responding crises and disaster management. In the new Bulgarian atlases of geography some maps of these themes have already been compiled.

If electronic variants of the atlases are produced, the updating and including of the new themes will be easy and faster. The main problem for countries like Bulgaria is the limited availability of computers in geography and history classes in schools. The computer labs are occupied by the lessons of electronic sciences and informatics. Teachers do not have the opportunity to use them, but pupils could use such atlases for homework and studying.

To help teachers in difficult topics, cartographers could be ready to supply them with additional materials, maps, and other cartographic products. Such example is

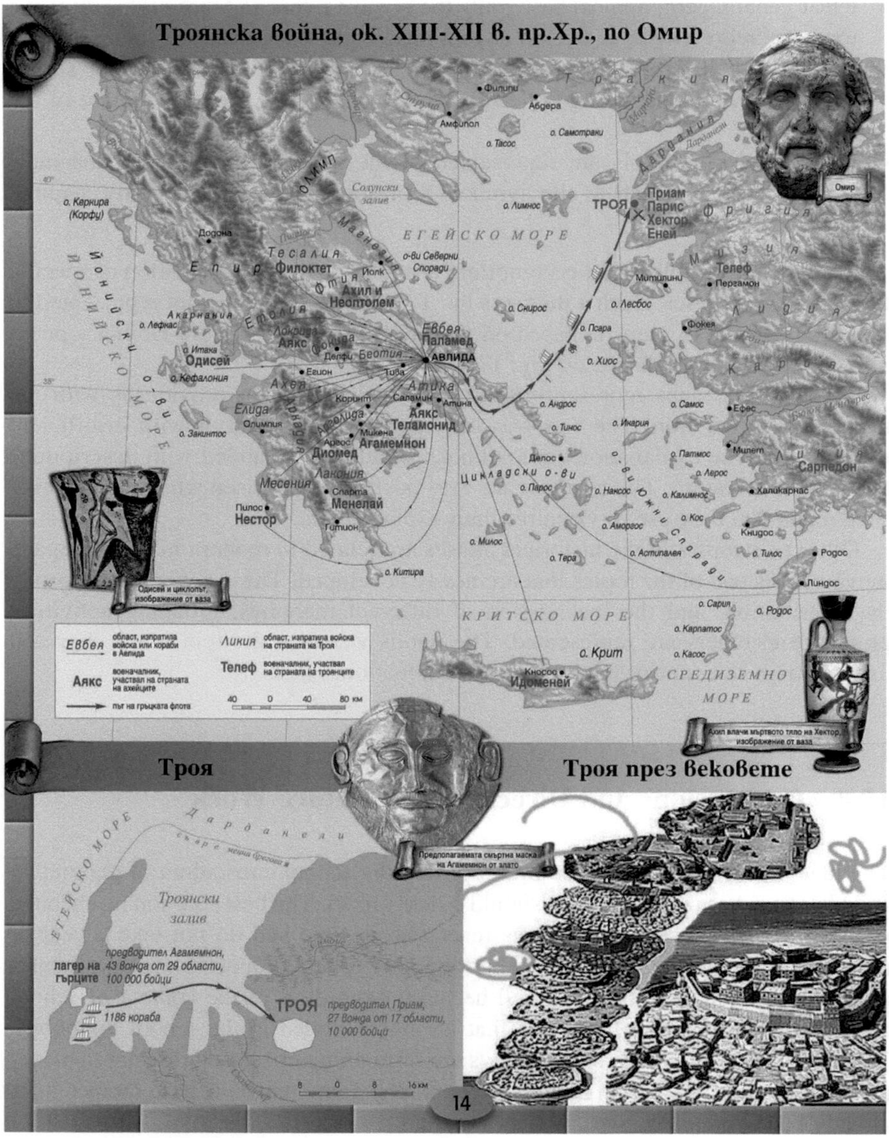


**Fig. 17.9** A map of the Trojan war published in the historical atlas for school year 7

an icosahedron product, which will help the teachers when they explain the topic of map projections. Some new ideas for the visualization of topics that were not proposed up to now will also help teachers and pupils in their educational process (see the map of the Trojan War).

The use of modern visualization methods in the new cartographic products (paper and digital) will be the cartographers' target to reach the pupils' attention and help them better understand the topics. The electronic versions of atlases and maps give the cartographers most power tools to do this by animation, sound and videos, and 3D presentations. This also allows the users to produce their own products. This will be our tasks in the very close future to supply schools, teachers and pupils with new cartographical products. This will be our response to school curriculum changes.

## References

Anderson J, Carriere C, LeSann J (2003) A pilot electronic school atlas of Quebec. Int Res Geogr Environ Educ 12(4):383–390

Bandrova T (2001) The first Bulgarian electronic atlas—a possibility for cartographic teaching in schools. E-mail Seminar of Cartography, 3:93–98

Bandrova T (2007) Cartography—close to users: a technology for compiling of school atlases for education on geography in Bulgaria. In: Proceedings of the XXIII international cartographic conference, Moscow

Bandrova T (2010) Bulgarian cartography: from paper to virtual reality. In: Proceedings of the 3rd international conference on cartography and GIS, Nessebar

Gold C (2008) Cartographic education in Canada. http://www.voronoi.com/wiki/images/6/66/Cartographic_education_in_canada.pdf. Accessed 09 Mar 2011

Konecny M, Svancara J (1996) A perception of the maps by Czech school children. In: Proceedings of the ICA seminar on children and education in cartography, Gifu, pp 137–146

McKibben B (2011) Natural disasters? The guardian online. http://www.guardian.co.uk/world/2011/apr/02/natural-disasters-floods-earthquakes-landslides. Accessed 03 Apr 2011

Ormeling F (1996) Teaching map use concepts to children. In: Proceedings of the ICA seminar on cognitive map, children and education in cartography, Gifu

Paris SG, Byrne JP (1989) The constructivis approach to self-regulation and learning in the classroom. In: Zimmerman BJ, Schunk H (eds) Self-regulated learning and academic achievement: theory, research and practice. Springer, New York, pp 169–200

Pfander J, Kollen C, Greenfield L (2004) The geographic literate student: the Arizona electronic atlas experiment. http://proceedings.esri.com/library/userconf/educ04/papers/pap5050.pdf. Accessed 09 Mar 2011

Reyes Nunez J, de Moretti C, Garra A, Erika G, Rey C, de Castro V, Dibiase A (2005) Reading thematic maps in Argentine and Hungarian schools: experiences in both countries. In: Proceedings of ICA ICC 2005 mapping approaches into a changing world, A Coruna

Reyes Nunez J, de Moretti C, Garra A, Erika G, Rey C, de Castro V, Dibiase A (2008) Resuming an international project: map use in Argentine and Hungarian schools. In: Proceedings of the second international conference on cartography & GIS, Borovets, pp 113–122

Stoltman J (1992) Geographic education in the United States: the renaissance of the 1980s. In: Naish M (ed) Geography and education. Institute of Education, London University, London

The New York Times (2011) Haiti, overview, world. http://topics.nytimes.com/top/news/international/countriesandterritories/haiti/index.html. Accessed 03 Apr 2011

Uluğtekin N, Bildirici İ (1997) Advanced mapping technologies: electronic atlases. In: Second Turkish-German joint geodetic days, Berlin, pp 641–649

Wortley D (2008) GIS and virtual worlds. The importance of place and community. In: Bandrova T (ed) Second international conference on cartography and GIS, UACEG, Sofia, vol 1, pp 317–322

Zhang H, Foskett N (2003) Changes in the subject matter of geography textbooks: 1907–1993. Int Res Geogr Environ Educ 12(4):312–329

# Chapter 18
# Research on Cartography for School Children

**Rosangela Doin de Almeida**

**Abstract** The aim of this article is to present research about cartography for school children carried out in the Research Laboratory for the teaching of Geography and Cartography (LABENCARTOGEO) of the Postgraduate Program in Geography of the Geoscience and Exact Sciences Institute of the State University of São Paulo (Rio Claro Campus, São Paulo, Brazil). This laboratory was created/founded in 2005 with the objective of carrying out research with teachers and schools, based on the Qualitative Research of Education approach. Since 1996, we have conducted theoretical studies using this methodology. However, in this article we are going to include information from previous studies so that the reader can follow our line of approach. We are going to present you with some background information regarding studies of teaching maps to children in Brazil, with the purpose of embedding the themes that we investigated in our research framework. At the end we will recommend future directions for the research to be conducted by researchers from LABENCARTOGEO.

## 18.1 Introduction

When LABENCARTOGEO was founded, two pieces of research were carried out on the base of Jean Piaget's (Piaget and Inhelder 1993) learning theory and his recommendations for the teaching of maps, which will be mentioned at a later stage. We developed three extensive studies from 1998 to 2008 as to how to produce local school atlases and the professional development for teachers. More recently, we have been working with multimedia cartography and the Internet and also with time and space references in kindergarten and pre-school education. We intend to present the methodology used and the results of all this research.

R.D. de Almeida (✉)
Universidade Estadual Paulista Julio de Mesquita Filho—UNESP, Rio de Janeiro, Brazil
e-mail: rda.doin@gmail.com; rdoin@rc.unesp.br

L. Zentai and J.R. Nunez (eds.), *Maps for the Future*,
Lecture Notes in Geoinformation and Cartography 5,
DOI 10.1007/978-3-642-19522-8_18, © Springer-Verlag Berlin Heidelberg 2012

## 18.2 Research Based on the Jean Piaget's Approach

The impact of Jean Piaget's theory in Brazilian education was very strong throughout the 70s and 80s. Regarding the teaching of maps, Lívia de Oliveira's (1978) thesis influenced studies in this area and served as a reference for various research papers afterwards. This paper was the first to present a detailed study of Piaget's ideas regarding a child's representation of space. In accordance with this line of studies, we published a book, which has been used by teachers for teaching maps (Almeida and Passini 1989). We based our thesis (Almeida 1994) on Piaget's ideas, which lead to other publications about cartographic concepts for primary school children (Almeida 2002, 2007).

In order to continue this approach, we supervised Miranda's research (2001) on the construction of a contour line concept by students of 11 and 12 years of age through observations in the field and of a model done to scale. The results showed us that the students are capable of identifying contour lines on large-scale maps (1:1,000) and indicating which way the water should drain. However, they had some difficulties in drawing contour lines on hydrographical maps and landmarks they are familiar with.

Another study which followed the same approach was carried out by Valéria Cazetta (2002) and dealt with the concept of land use based on aerial photographs. She used aerial photographs of the city's neighborhoods with students of 12 and 13 years of age (scale 1:5,000) to produce layers with a pre-defined key. In order to identify the urban areas (built-up areas, uninhabited areas, squares and areas occupied by large buildings). The research showed that the students had no difficulties constructing the layers, however not all of them were able to identify different types of neighborhoods based on the designated land use. We concluded that additional knowledge is necessary for this concept to be learned, for example, learning the historical background of the city.

We consider the Piagetian approach suitable for investigating the acquisition of concepts related to spatial relations (topological, projective and Euclidian) and to the logical structure of thought, however there are other factors, which interfere with the learning process such as symbolic exchanges (through speech, gestures, written forms, visual images, etc.) and the factors in the institutional and social context. Thus we attempted other theories for producing school atlases with the collaboration of the teachers. We changed the way we approached the projects on the basis of the premise that *time* interferes in the production of knowledge. It is necessary to allocate more time to the discussions so that there is interlocution between the researchers and teachers (Bakhtin 1981). It is in the interlocution that new knowledge arises and this is a methodological condition for the nature of the research that we do. Another fundamental point is language, as knowledge and language go hand in hand in terms of understanding the issues related to the representation of space.

## 18.3 Other Approaches for Research

A new approach enabled us to attain the objectives that we had defined related to the *curriculum*, *school culture* and the *teacher's qualifications*. We mainly based our research of the curriculum and school culture on Ivor Goodson's (2001) ideas. He considered the curriculum to be a result of the power plays between the dominant groups within higher educational institutions and the social lobby groups that have a direct influence on schools and the community. This enabled us to observe how effectively this cartographic knowledge is implemented in the school's curriculum. We are aware that in the educational process that the teachers are the most important agents, therefore we observed the teacher's practices in the classroom. To this end we used L. Stenhouse's, John Elliott's and Kenneth M. Zeichner's publications presented by Corinta Geraldi (ed 1998). Other authors that were important in shaping our research methodology were D. Jean Clandinin and F. Michael Connelly (1995, 2000).

This theoretical approach confirmed various studies related to the production of local school atlases at Master and Doctorate levels, as is shown in the table above:

| Date | Level | Title | Short abstract |
|---|---|---|---|
| 2002 | M | The production of the municipal atlas of Santa Maria (RS) | The greatest difficulty faced by teachers attempting to produce a local Atlas was their lack of knowledge of cartographic visualization that is adequate for primary school students. This resulted in maps that were confusing, inaccurate and overly detailed for the students (Viero 2002) |
| 2004 | M | A history of Ipeúna—SP | As this city didn't possess historical records, the researcher put together a historical account for the School Atlas based on the methodology of research using oral history and statements of former residents (Machado 2004) |
| 2005 | D | Educational practices, mapping processes and aerial photographs: Passages and constitution of knowledge | The researcher presents cartography as passage between human experiences in space and a search for the production of sense (Cazetta 2005) |
| 2006 | M | Challenges that arose in the teaching-learning process of place in the early primary years: Possibilities to achieve citizenship | The researcher observes and analyses a series of classes of one group of students whose families had moved (migrated) a number of times. Use of the school atlas during classes had a certain impact on the comprehension of local space and its representation, once the neighborhood where the school was located a long way from the city. This made it difficult for the |

(continued)

| Date | Level | Title | Short abstract |
|---|---|---|---|
| | | | students to comprehend the city and its cartographic representation (Santos 2006) |
| 2006 | D | The official curriculum and the curriculum practiced by teachers of a particular network | The researcher discussed the teacher's practices who participated in this networking group. The focus of the research was the material dimensions and symbols of the idea of "place" (Gonçalves 2006) |
| 2006 | D | The connection between orality and visuality: reading the world through maps | The researcher discusses the connection between orality and visuality on the thematic maps in the Atlas of Rio Claro. She uses Deleuze's theory of mediation and the presuppositions of Mikhail Bahktin (1981) and Lev Vygotsky (1988) as references (Aguiar 2006) |
| 2007 | D | Establishing fundamental Theories for the Internet Atlas Executed and applied in the Brazilian Primary Education System [Held at RMIT University, Melbourne, Au, under the supervision of William Cartwright] | This research uses the school atlas of Rio Claro (Sao Paulo State, Brazil) as a case study. The original files used for the production of the paper atlas were evaluated and then procedures for converting them to an SVG product were developed. To achieve this goal, different graphical packages were evaluated and an interactive prototype was built in using JavaScript. The interface was developed using simple colors and basic shapes, so that the final template can be implemented by people with a basic knowledge of graphic design, and little prior knowledge of SVG. The online atlas prototype was implemented and tested in real class situations in Brazil (Ramos 2007) |
| 2008 | M | Here, there and everywhere: ways and experiences of teachers in teaching of place | Upon looking at ethnographic records of the classes of three teachers the researcher realized that knowledge and mainly the involvement of the teacher with the place where that teach interferes to a great extent in their classes. The use of a local atlas doesn't change the teacher's way of teaching, rather the notion that the teacher has of the place (Camargo 2008) |

(continued)

| Date | Level | Title | Short abstract |
|---|---|---|---|
| 2008 | M | Teacher's practices in terms of place and school cartography in the context of collaborative research | This study discusses how the teachers that participated in this group of collaborative research brought about an exchange and production of knowledge in terms of teaching about place through the means of a local atlas. It was concluded that the change in their practices could not be obtained without interacting with the research group<br>The teachers learn to use the maps and photos of the atlas as a way to teach geographical concepts, which they didn't do adequately before the study (Locali 2008) |

Based on this body of research we reached the following conclusions about teaching with a local atlas and the fallout of a collaborative group:

- The teachers' knowledge and practices about *place* are not composed of a clearly defined and continuous space. Their knowledge is composed of fragments of space and time, mediated by physical and symbolic elements with which they come into contact, so are fragmented too.
- Immigrants and the media give rise to "place" a cultural diversity, leading to a historical discontinuity between the concepts of social groups who attend the school. This causes serious difficulties when teachers are focused only on the teaching of geography and historical development of the region without taking into account the cultural references of the students.
- The local Atlases should be less standardized and more open to the inclusion of cartographic representations with content related to the culture of the students and the teaching practices of teachers.

Another important topic for the study of School Cartography is related to the origins of the content of cartography in the school curriculum. Based on the ideas of Ivor Goodson (2001), a study was conducted of the didactic Geography books published from 1824 to 1940. In this study, concluded in 2010, we discovered that a set of notions, concepts and topics, such as "Direction and Orientation", "Shape of the Earth" and the "Movements of the Stars", "Imaginary lines: Parallels and Meridians", "Geographic coordinates/Latitude and Longitude", "Map" and "Globe" are items that have remained in the curriculum throughout the last two centuries (Boligian 2010). Besides this, they are taught in nearly the same way even today. Even though teachers today use advanced resources, the teaching method remains quite descriptive.

Currently, we are carrying out investigations of the topic *Multimedia Cartography, Internet and Knowledge*. In 2010, research about *Cartography in the Cyber Culture Age: mapping other types of Geography in Cyberspace* (Canto 2010) was completed. Mapping projects were analyzed using new programs that are available on the Web. These resources permit the users to obtain maps in the following ways:

- The users took advantage of the open, interactive and global quality of this new space of communication to construct collaborative representations or map out personal stories.
- As these systems allow one to surf the Earth virtually, users create spatial relations that are only possible on the Internet with the real places they know.
- Users create spatial relations with places that they got to know in in the virtual media. Thus, the new term of space-time made the map an instrument capable of presenting new realities.

One of the sites analyzed was the Urban Post (http://post.wokitoki.org) created by Daniel Perosio in Rosário, Argentina. This site is the result of a combination of Google Maps with other tools for the production of collaborative mapping. The objective of the project was to show to the inhabitants and visitors areas hidden by the city's planning. By means of notes—*posts*—added to satellite images and maps, the people would describe their experiences in the city and afterwards posters with these messages were put up in the areas of the city that marked that site.

Another two studies were carried out in 2010 about cartography and the Internet. One of them dealt with the concept of interactivity in multimedia cartography projects. The term multimedia was incorporated into cartography and is used for a combination between cartographic representations with other forms of media such as texts, figures, videos, sound and animations (Peterson et al. 1999). The word "*interactivity*" was created to emphasize a qualitative change in relation to the user with computer interfaces, when devices that allow for the input and output of data were incorporated in computational systems. These allow for interaction between man and machine. With the purpose of analyzing multimedia cartography projects, the researcher (Gomes 2010) created the following classification of level of interactivity:

| Category | Levels of interactivity | General characteristics |
|---|---|---|
| *Interactivity of animation* | *Level 01* | It is not permitted to interfere in the sequence of the presentation (stop, interfere, return), nor the interaction through means of any modification of any variable that alters the simulation. The interactivity is restricted to the option of repeating the presentation |
| | *Level 02* | It is permitted to interfere in the presentation (stop, continue, return, repeat) through means of manipulation of variables that may alter the simulation, enabling the visualization of other situations foreseen by the creator of the project |
| *Interactivity of selection* | *Level 01* | This interactivity is limited to the option of choice before stopping, continuing or going back, following an order of *linear* movement according to the content, within a certain hierarchical framework |
| | *Level 02* | Allows the *linearity* of movement to be broken by the project, making it possible for the *user* to choose the content which he/she wishes to access in a *non-linear* way |

(continued)

| | | |
|---|---|---|
| *Interactivity of recreation* | *Level 01* | This allows for the comparison of items, through means of simulation, using preexisting information, in accordance with an individual characteristic and the necessity of the *user* |
| | *Level 02* | This allows for updating information, modifying the content and recomposing the message as needed, making co-authoring possible |

Other research that was completed last year looked at the concept of interactivity in cartographic multimedia projects and the use of Google Maps as a didactic resource for mapping local space for children from 11 to 12 years of age (Fonseca 2010). Both studies look at *interactivity*, which is the fundamental concept for studies regarding mapping in multimedia projects that involve time, space and culture.

We have recently begun a study looking at pre-school children's representation of space. The objective is to put together a collection of files recording teaching situations with children of 3 and 4 years of age. This collection enables us to analyze how children in this age-group deal with situations that mobilize relations between time-space-body.

## 18.4 Conclusions

At the start of the article we mentioned that we are carrying out our research based on the idea of *school culture*. In a nutshell, knowledge and school practices are considered to be *social construction* and not as knowledge originating from a prescribed curriculum. This knowledge is loaded with cultural values coming from the education institution itself. The power relationships exercised in the school context modify the traditional hierarchy between student/knowledge/teacher. In the context of school culture knowledge becomes a social construction and the method of teaching is affected by the influences originating outside the school. Even though learning is a personal construction, there are many concepts that are common because they are formed by the interaction of personal experiences in the same cultural context. The school is seen as a scenario where the students develop their personal experiences, allowing them to re-construct and construct them together with new knowledge. Different language types (oral, written, graphic, imagery. . .) allow for this construction, and the construction of social-spatial knowledge is achieved through *cartographic language*.

## References

Aguiar LMB (2006) A cumplicidade entre a oralidade e a visualidade: lendo o mundo através dos mapas. Dissertation, Instituto de Geociências e Ciências Exatas da Universidade Estadual Paulista, São Paulo

Almeida RD (1994) Uma proposta metodológica para a compreensão de mapas geográficos. Dissertation, Universidade de São Paulo, Faculdade de Educação. São Paulo
Almeida RD (2002) Do desenho ao mapa, Iniciação cartográfica na escola. Contexto, São Paulo
Almeida RD (ed) (2007) Cartografia escolar. Contexto, São Paulo
Almeida RD, Passini EY (1989) O espaço geográfico: ensino e representação. Contexto, São Paulo
Bakhtin M (1981) Marxismo e filosofia da linguagem. Hucitec, São Paulo
Boligian L (2010) Materiais escolares, imperativos didáticos e currículo de Geografia: contribuições para a história da Cartografia Escolar no Brasil. Dissertation, Universidade Estadual Paulista, Instituto de Geociências e Ciências Exatas, São Paulo
Camargo PEB (2008) Aqui, Ali e Acolá: Caminhos e Experiências do Lugar em Práticas Docentes. Dissertation, Universidade Estadual Paulista, Instituto de Geociências e Ciências Exatas, São Paulo
Canto TS (2010) A cartografia na era da cibercultura: mapeando outras geografias no ciberespaço. Dissertation, Universidade Estadual Paulista, Instituto de Geociências e Ciências Exatas, São Paulo
Cazetta V (2002) A aprendizagem escolar do conceito de uso do território por meio de croquis e fotografias aéreas verticais. Dissertation, Universidade Estadual Paulista. Instituto de Geociências e Ciências Exatas, São Paulo
Cazetta V (2005) Práticas educativas, processos de mapeamento e fotografias aéreas verticais: Passagens e constituição de saberes. Dissertation, Universidade Estadual Paulista, Instituto de Geociências e Ciências Exatas, São Paulo
Clandinin DJ, Connelly FM (1995) Relatos de Experiencia e Investigación Narrativa. In: Larrosa J et al (eds) Déjame que te Cuente: ensayos sobre narrativa y educación. Alertes, Barcelona, pp 11–59
Clandinin DJ, Connelly FM (2000) Narrative inquiry. Experience and story in qualitative research. Jossey-Bass, San Francisco
Fonseca RA (2010) Uso do Google Mapas como recurso didático para mapeamento do espaço local por crianças do Ensino Fundamental da cidade de Ouro Fino (MG). Dissertation, Universidade Estadual Paulista, Instituto de Geociências e Ciências Exatas, São Paulo
Geraldi CMG (ed) (1998) Cartografias do trabalho docente. Mercado de Letras, Campinas
Gomes S (2010) Cartografia Multimidia: interatividade em projetos cartográficos. Dissertation, Universidade Estadual Paulista, Instituto de Geociências e Ciências Exatas, São Paulo
Gonçalves AR (2006) Os espaços-tempos cotidianos na Geografia Escolar: do currículo oficial e do currículo praticado. Dissertation, Universidade Estadual Paulista, Instituto de Geociências e Ciências Exatas, São Paulo
Goodson I (2001) O currículo em mudança. Estudos na formação social do currículo. Porto, Portugal
Locali R (2008) Práticas Docentes Sobre Ensino do Lugar e Cartografia Escolar no Contexto de uma Pesquisa Colaborativa: Processos de uma Cosntrução. Dissertation, Universidade Estadual Paulista. Instituto de Geociências e Ciências Exatas, São Paulo
Machado HMFG (2004) Uma história para Ipeúna (SP). Dissertation, Universidade Estadual Paulista, Instituto de Geociências e Ciências Exatas, São Paulo
Miranda SL (2001) A noção de curva de nível no modelo tridimensional. Dissertation, Universidade Estadual Paulista, Instituto de Geociências e Ciências Exatas, São Paulo
Oliveira L (1978) Estudo metodológico e cognitivo do mapa. In: Série Teses e Monografias, 32. Universidade Estadual Paulista, Instituto de Geociências e Ciências Exatas, São Paulo
Peterson M, Cartwright W, Gartner G (eds) (1999) Multimedia cartography, 2nd edn. Springer, Berlin
Piaget J, Inhelder B (1993) A representação do espaço na criança (trans:. de Albuquerque BM). Artes Médicas, Porto Alegre
Ramos CS (2007) Establishing Fundamental Theories for Internet Atlas Realisation with Application in the Brazilian Primary Education System. Dissertation, RMIT University, Royal Melbourne Institute of Technology, Melbourne

Santos GA (2006) Desafios no processo ensino-aprendizagem do lugar nas séries iniciais do Ensino Fundamental: possibilidades para a formaçao para a cidadania. Dissertation, Universidade Estadual Paulista, Instituto de Geociências e Ciências Exatas, São Paulo

Viero LMD (2002) A elaboração de um atlas escolar municipal como uma contribuição para o ensino de Geografia- Santa Maria (RS). Dissertation, Universidade Estadual Paulista, Instituto de Geociências e Ciências Exatas, São Paulo

Vygotsky LS, Luria AR, Leontiev AN (1988) Linguagem, desenvolvimento e aprendizagem. Ícone, São Paulo

# Chapter 19
# The World in Their Minds: A Multi-scale Approach of Children's Representations of Geographical Space

**Veerle Vandelacluze**

**Abstract** A number of studies have explored primary school children's spatial representations at one particular scale. Little is known however on children's mental constructs of the geographical space as a whole. The omnipresence of digital mapping tools in everyday life has made such an understanding increasingly important. Elaborating effective teaching strategies e.g. for Google Earth or GPS, requires a basic understanding of children's mental maps at different scales. In this study 94 Flemish children aged 9–12 represented their world at three scale levels: the local neighbourhood (test 1), the continents (test 2) and the Sun-Earth-Moon system (test 3). The results demonstrate that a majority of the children show an incomplete geographical world view. Only between test 2 and 3 a significant correlation was found. This supports previous psychological research pointing out at least a partial dissociation between large-scale and small-scale spatial abilities. Consequences for primary school map education are discussed.

## 19.1 Introduction

The omnipresence of digital mapping tools in everyday life asks for an introduction of visualization tools like Google Earth or GPS into the primary school classroom. The elaboration of effective teaching strategies to do so requires a profound understanding of children's mental maps and spatial abilities at different scales. During the last decades a number of studies have explored primary school children's spatial representations at one particular scale e.g. earth's land masses or the familiar home-school area. Little is known however on children's mental constructs of the geographical space as a whole.

V. Vandelacluze (✉)
Department of Primary School Teacher Education, University College KATHO—Association Catholic University of Leuven, Tielt, Belgium
e-mail: veerle.vandelacluze@katho.be

L. Zentai and J.R. Nunez (eds.), *Maps for the Future*,
Lecture Notes in Geoinformation and Cartography 5,
DOI 10.1007/978-3-642-19522-8_19, 

This research aims to contribute towards a broader and more holistic understanding on what the world inside children's heads looks like through exploring their representations and spatial abilities at three different geographical scales: the local neighbourhood (test 1), the continents (test 2) and the Sun-Earth system (test 3).

## 19.2 Theoretical Background

### *19.2.1 The Psychology of Geographical Space*

When assessing children's mental maps of the world based on physical representations, it is necessary to differ between psychological and physical space. Because we are dealing with different spaces and scales, it is at the beginning of this paper also important to "recognize the distinction between geographical space and space at other scales or sizes" (Mark et al. 1999). Besides defining geographical space, the researched space in this study, its position within this psychological-physical space duality needs to be addressed.

In 1978 O'Keefe and Nadel defined *psychological space* as "any space that is attributed to the mind (...) and which would not exist if minds did not exist" and *physical space* as "any space attributed to the external world independent of the existence of minds". Many researchers have referred to this definitions since. While children's mental maps belong to psychological space, the outcome measures (external expressions of mental maps) used in this research respectively the route map, the composed map of the continents and the Sun-Earth-Moon scheme are *in se* physical space.

The precise nature of *geographical space* on the other hand is not that easy to describe. Geographical space being the space "... generally too large to be perceived all at once" is an often referred to description provided by Downs and Stea (1977). Building on this Mark et al. (1999) elaborated it and noted that while geographical or large-scale spaces can only be experienced by "integration of perceptual experiences over space and time through memory and reasoning, or through the use of small-scale models such as maps", small-scale spaces "are to be seen from a single point, and typically are populated with manipulative objects, many of which are made by humans". To avoid terminological confusion it is to be noticed here that for behavioural scientists large-scale spaces are relatively large compared to small-scale spaces. Opposite, for geographers "a small-scale map is small compared to the space that it represents" (Montello 1993). Therefore, when focussing on spatial abilities (*see further* Large- and small-scale spatial abilities) the former interpretation, when referring to the children's representations of the geographical space at different scales, the latter interpretation will be used in this paper. The children's route maps (test 1) are thus large-scale compared with their map of continents composition (test 2) and their Sun-Earth-Moon reproduction (test 3).

On the face of it geographical space obviously holds a physical dimension as in landscape elements e.g. trees, buildings, roads... However, the required cognitive structures or mental maps needed to understand spatial relationships are to be located in psychological space. Still, question remains where exactly to locate geographical space within psychological space. Remark in that respect that in both attempts to describe geographical space mentioned above (Downs and Stea 1977; Mark et al. 1999) scale naturally seems to appear. Montello (1993) indeed found that scale has an important influence on how humans treat spatial information and elaborated, building on previous research, a categorization of psychological space especially useful for this research because it contains "geographical space" as one of the four categories. He argues that space is not scale independent and therefore bases the categorization on scale differences. His four categories depending on a person's point of view are:

1. *Figural space*—smaller than the body and can be subdivided in
    - Pictorial space, small flat spaces
    - Object space, small 3D spaces
2. *Vista space*—larger than the body but visually apprehensive from a single place
3. *Environmental space*—surrounding the body, too large to apprehend without considerable locomotion and integration of information over significant periods of time
4. *Geographical space*—much larger than the body, not apprehensive through locomotion but via symbolic representations such as maps or models

Attention must be drawn to the fact that although maps are representations of environmental and geographical spaces, they themselves are part of the figural space considering maps can be studied without moving. Returning now to the question where to position geographical space, conclusion appears to be that although it is psychological and physical the largest space, it can't be mentally structured without small symbolic representations (physical space) inherent to the psychological figural space such as maps, aerials ... (pictorial space) or 3D-models (object space).

Because learnt through daily travel experience and maps or images of the neighbourhood, children's route maps from home to school (test 1) contain an environmental and a figural spatial dimension. Therefore, assuming that by the age of nine children will have explored a map or aerial, the mental maps—from which the route maps are derivatives—are placed in the geographical space as well as in the environmental space or better still in the transition zone between the environmental and geographical space. The construction of a coherent mental map of the world (test 2) typically educated through maps and globes obviously concerns geographical space. When developing a correct image of the Sun-Earth-Moon system (test 3) one could argue that we reach the ends, if they already existed, of

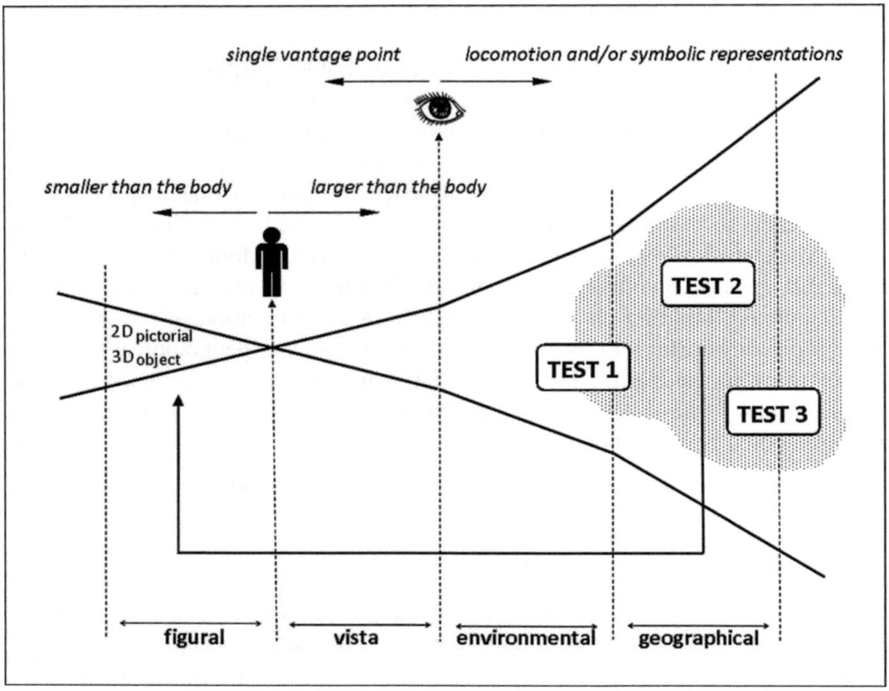


**Fig. 19.1** Positioning geographical space in Montello's classification of psychological spaces

the geographical space entering what can be called "cosmological space" (*see* Fig. 19.1).

The assessed spatial representations in this study are thus the physical outcomes of the children's cognitive maps of geographical space. These mental constructs however are not purely geographical *in se* (since they are related to different areas of psychological space as explained above) but merely geographical concepts internalized in the whole of a child's psychological space.

### *19.2.2 Developing Cognitive Models and Spatial Abilities at Different Scales*

Although by the time they enter school most children have already adopted a range of ideas about their geographical surroundings and the wider world, these notions tend to be rather confused (Scoffham 1999). Even very young children do create cognitive maps. As the child matures its cognitive maps and ability to represent them evolve. Children up to 6 years old generally have egocentric spatial understandings. Their images become more objective around the age of 7. By the age of 10 the child develops abilities that cope with abstract spatial knowledge and concepts (Catling 1979).

It is therefore reliable that the children in this study (aged 9–12) are able to represent their mental maps at different geographical scales and show the beginning of higher spatial thinking ability needed to complete the three tests.

Small-scale spatial abilities are used to perform "small-scale" spatial tasks at the *figural* scale of space (smaller than the body, *see* Fig. 19.1) e.g. for imagining or mentally transforming small shapes. Large-scale spatial abilities on the other hand are needed to perform *environmental* spatial tasks, such as way-finding or learning the layout of spaces that surround the body and involve integration of the sequence of views that change with one's movement in that environment (Hegarty et al. 2006).

Most research has focused on small-scale psychometric tests such as mental rotation of shapes, finding hidden figures, or imagining the folding and unfolding of sheets of paper. There have been relatively few attempts to assess spatial ability in larger spaces such as the geographical space, the researched space in this study. In their attempt to do so Quaiser-Pohl et al. (2004) found a distinction between large-scale tasks, where the observer is part of the environment and cannot see the whole space of interest at once and small-scale spatial tasks, where spatial relations of objects can be seen at once. Hegarty et al. (2006) conclude on the basis of previous extensive literature review and their own test results and analysis that there are indications that spatial abilities at different scales of space are *partially but not totally dissociated*.

> . . .it is likely that the ability to remember the sequence of landmarks along a route does not share common processes with small-scale spatial cognition, but the ability to infer the configuration of an environment from route experience does. (Hegarty et al. 2006)

Thus, it is assumable that when applying landmark strategies to draw a route map of a familiar environment we use our large-scale spatial abilities, but when internalizing the configuration of that environment or to apprehend larger spaces that cannot be learnt by route experience we demand more up on our small-scale spatial abilities (*see* Fig. 19.2).

Siegel and White (1975) suggested that children first master landmarks then routes and later on develop the ability to configure their environment. A sequence that corresponds with the developmental stages suggested by Piaget and Inhelder (1956) ". . .insofar as the use of landmarks is akin to the use of *topological* concepts (e.g., knowing that the school is next to the supermarket), the use of routes is akin to the use of *projective* concepts (e.g., knowing that when traveling from home to school, one turns right at the traffic light, but on the return trip, one turns left at the light) and the use of survey knowledge is akin to the use of *Euclidean* concepts (e.g., conceptualizing locations of places and pathways by using metric distances and angles)" (Liben 2006).

Hence, children drawing the route from home to school (test 1) based on landmark strategies are expected to demand up on their large-scale spatial abilities. But children with a more evolved spatial cognition, who are, at least in some degree, able to conceptualize their local environment (Euclidean strategies) are likely to use

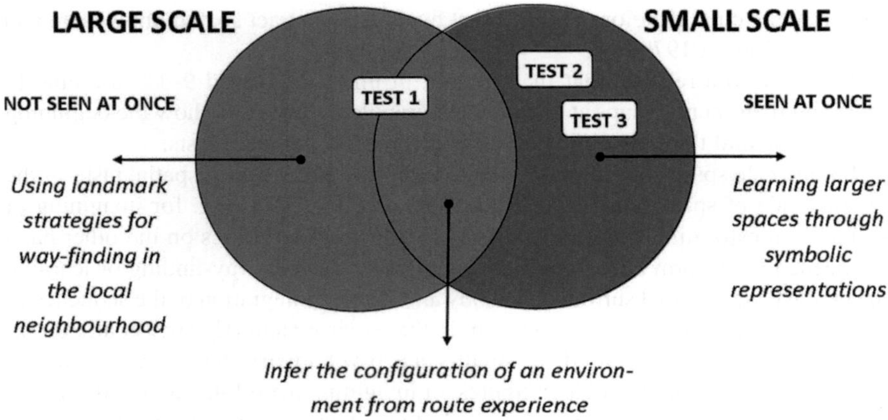


**Fig. 19.2** Relation between large- and small-scale spatial abilities

large- as well as/or small-scale spatial abilities when they draw their route from home to school. For test 2 and 3 on the other hand children would be forced to draw more exclusively up on their large-scale spatial abilities because landmark strategies could not help them completing this tasks.

Based on these indications of partial dissociation of large- and small-scale spatial abilities found in Hegarty's study cited above, it is thus more likely to find correlations between test 2 and 3 than with test 1. Children might draw a high quality route map of the local environment using landmark strategies (large-scale spatial abilities) but could perform much less on test 2 and 3 when their spatial cognition is less developed.

## 19.3 Methodology

The study was conducted in five Belgian primary schools, three in rural and two in an urban environment, spread over the Dutch speaking part of Belgium. 94 children (44 girls, 50 boys), aged 9–12 years, from five mixed ability classes were given approximately 30 min to complete the three tests (explained in detail in the following).

In order to achieve similar conditions in each class the test takers were given specific instructions such as "Do not help the children in any way"; "Only explain the task"; "Blank the map of the world if present on the classroom walls"... The children were motivated and stimulated to take the test seriously by means of a competition. The class with the best overall scores was promised a special prize (*see* Fig. 19.3).

The quality of the tests was measured via scoring tables. Each test was revised twice by independent persons (agreement >85% for all three tests). The mean scores were used for the statistical analysis (*see* 1.4 Results and discussion).

Fig. 19.3 Children completing the tests (*left*),winning class (*right*)

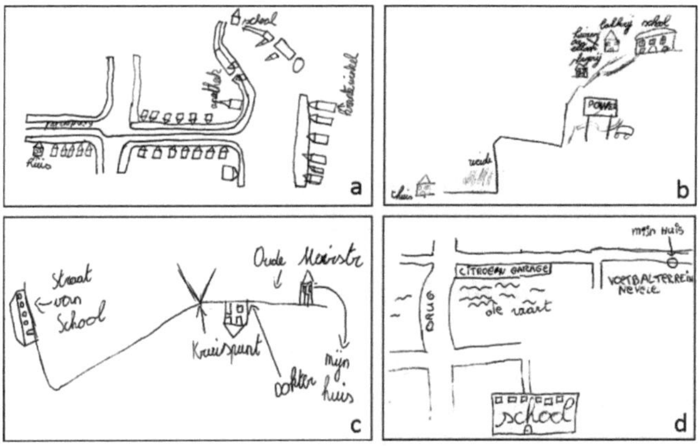


**Fig. 19.4** Robbe's (9), Lennert's (10), Lin's (11) and Amélie's (11) route

## *19.3.1 Test 1: The Route from Home to School*

Although free recall sketch maps have the methodological shortcoming of relying heavily on the subject's drawing ability (Siegel et al. 1978; Spicer 1984; Matthews 1992) this often used research method was selected for test 1 for a number of reasons: the task is easy to explain to the children, can be completed within the limited time available and assessed in different ways (meeting the aims of the research).

Children were asked to draw a map of their route from home to school. They were specifically asked to include features that are helpful to them in remembering the route and to draw the map like they see it in their head (Fig. 19.4).

The scoring table (*see* Table 19.1) for this test was constructed with the aim of giving a general indication of the quality of the represented space. To that end five different features were given a score (0, 1 or 2) expressing the quality of the representation (based on a study by Thommen et al. 2010). The five assessed features are: paths, crossroads, buildings, the presence of a legend and traffic signs.

**Table 19.1** Drawing the route from home to school (test 1)

| | |
|---|---|
| a. Paths | |
| 0 | Unclear directions |
| 1 | Single, curved lines or double lines with breakpoints |
| 2 | Single, curved lines or double lines with breakpoints and parts of the road not to be taken |
| b. Crossroads | |
| 0 | None |
| 1 | One type of cross-road |
| 2 | Two or more types of cross-roads |
| c. Buildings | |
| 0 | None |
| 1 | Home and/or school |
| 2 | At least one other building |
| d. Legend | |
| 0 | None or only home and school |
| 1 | Bus stop, friend's house, church, bridge, parking... |
| 2 | Names of streets and buildings |
| e. Traffic | |
| 0 | None |
| 1 | One type (e.g. zebra crossing, traffic light...) |
| 2 | Two or more types (e.g. zebra crossing, traffic light...) |

### *19.3.2 Test 2: Map of the Continents*

To compose a map of the continents (test 2) a technique similar to Wiegand and Stiell's (1996) cut-out continents was elaborated taking into account the methodological shortcomings of freehand sketch mapping mentioned above. Moreover, a limited preliminary research had shown a lack of motivation for sketch mapping a map of the world because of the difficulty of this task (which was not the case for sketch mapping the route from home to school).

The children were given a paper with the outline of the continents and asked to cut them out and paste them correctly (*see* Fig. 19.5).

For this test the scoring table (*see* Table 19.2) was based on previous studies (Wiegand 1995; Harwood and Rawlings 2001; Schmeinck 2007) and focuses on the location of the shapes, clusters and labels.

### *19.3.3 Test 3: The Sun-Earth-Moon System*

Previous research on children's understanding of the Earth as a cosmic body has shown that children perform considerably better on recognition than on recall tasks (Panagiotaki et al. 2006). Test 3 therefore comprised 2 tasks (*see* Fig. 19.6):

(a) Children had to mark out via arrows how Sun, Earth and Moon revolve around one another given three circles representing Sun, Earth and Moon;

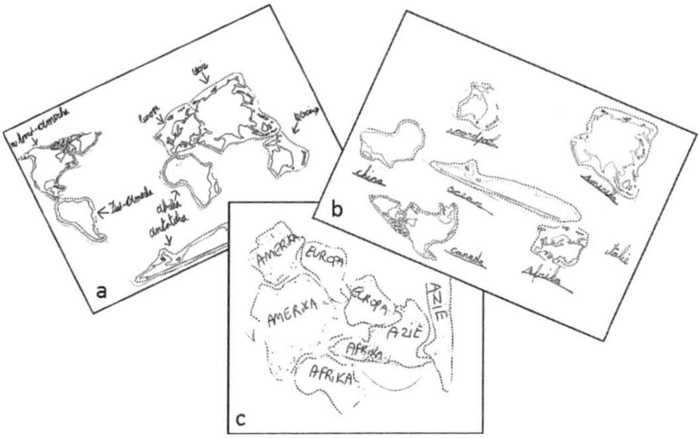


**Fig. 19.5** Bert's (11), Jelka's (9) and Raoul 's (12) composition

**Table 19.2** Composing a map of the continents (test 2)

| | |
|---|---|
| 1 | Isolated islands; no or less than three continents correctly labeled |
| | Randomly clusters; no or less than three continents correctly labeled |
| 2 | Isolated islands; three or more continents correctly labeled |
| 3 | At least two continents correctly clustered (label of outline correct) |
| 4 | Three continents correctly located; at least two continents correctly labeled |
| 5 | Four continents correctly located; at least three continents correctly labeled |
| 6 | Five continents correctly located; at least four continents correctly labeled |
| 7 | Six continents correctly located; at least five continents correctly labeled |
| 8 | Seven continents correctly located; at least six continents correctly labeled |

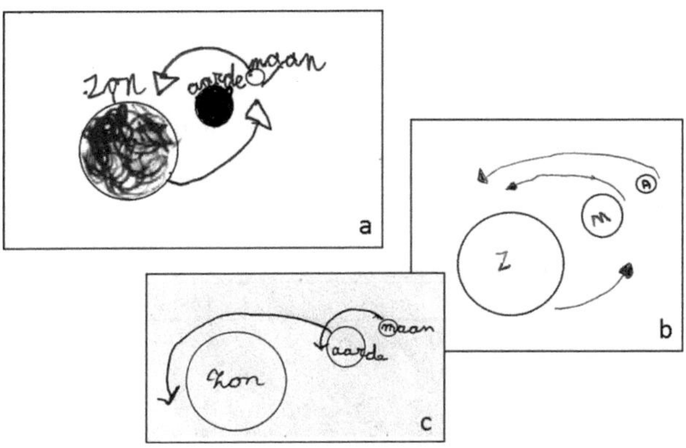


**Fig. 19.6** Jason's (10), Lotte's (11) and Sebastian's (10) scheme

(b) Children had to answer a multiple choice question offering six possibilities (children were asked to indicate all correct answers):

- The Sun revolves around the Earth
- The Earth revolves around the Sun
- The Moon revolves around the Earth
- The Earth revolves around the Moon
- The Moon revolves around the Sun
- The Sun revolves around the Moon

Because of the number of different conceptions (amongst others Samarapungavan et al. 1996; Roald and Mikalsen 2001) the distinction heliocentric/geocentric world view was central to the evaluation of test 3 (*see* Table 19.3).

## 19.4 Results and Discussion

Despite great individual differences the results of this sample demonstrate that a majority of the children, even among 11/12-year-olds, show incomplete mental maps at all three scale levels (only 7% achieved a high overall score). The girls scored slightly higher on test 1, the boys performed better on test 2 and 3. Children's performances on the three tests are roughly comparable with previous research in other countries on which the scoring tables were based (*see* Sect. 19.3). Because this paper focuses on overall performances at different spatial scales individual and group differences are not further discussed here.

**Table 19.3** Completing the Earth-Sun-Moon model and multiple choice question (test 3)

| | |
|---|---|
| 1 | Geocentric view |
| 2 | Heliocentric view: no, unclear or false arrows—some questions false but clear heliocentric view |
| 3 | Heliocentric view: no arrows/questions correct or correct arrows/some questions false |
| 4 | Heliocentric view: correct scheme and answers |

**Table 19.4** Correlations between test 2 and 3

| | | Test 2 | Test 3 |
|---|---|---|---|
| Test 2 | Spearman rho | 1 | .371[a] |
| | Sig. (2-tailed) | | .000 |
| | N | 94 | 94 |
| Test 3 | Spearman rho | .371[a] | 1 |
| | Sig. (2-tailed) | .000 | |
| | N | 94 | 94 |

[a]Correlation is significant at the 0.01 level (2-tailed)

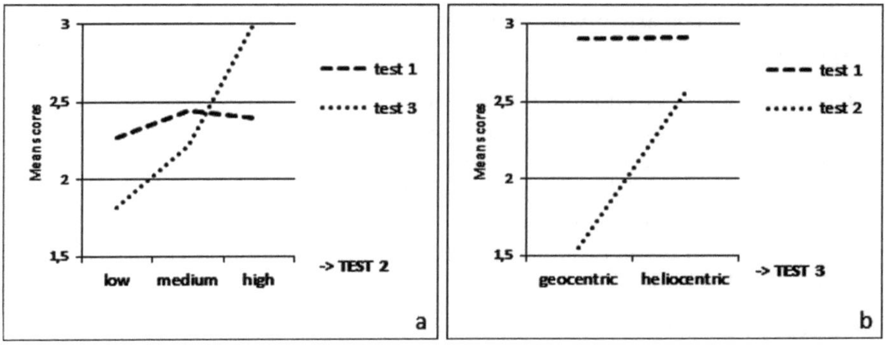


**Fig. 19.7** Relationships between the test results

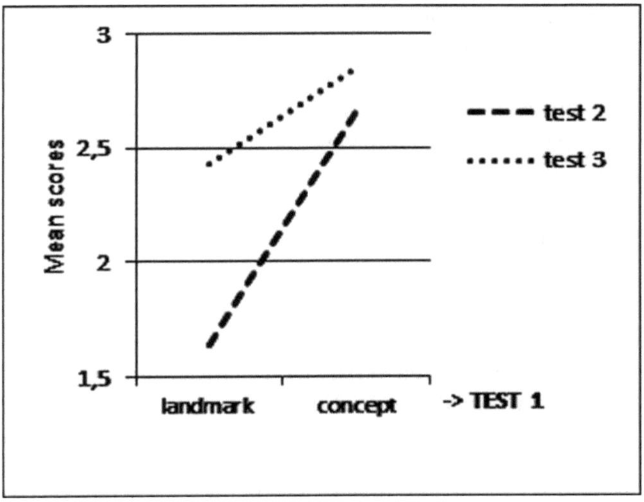


**Fig. 19.8** Landmark versus Euclidean strategies

Spearman's rank correlation was used to assess possible relationships between children's performances at the three spatial scales. Only between test 2 and 3 a significant correlation was found (rho = 0.37; $p < .000$) (*see* Table 19.4). Children scored similar (low, medium or high) on test 2 and test 3 (*see* Fig. 19.7a). Children showing a geocentric geographical world view on test 3 obtained lower scores on test 2 than the children who expressed a heliocentric view on test 3 (*see* Fig. 19.7b).

Half of the children (age 9–10: 66%; age 11–12: 41%) clearly made use of landmark strategies on test 1. Approximately 25%, using Euclidean strategies, were able to conceptualize the local environment (*see* Fig. 19.4a, d). Those children obtained higher scores on test 2 and 3 (*see* Fig. 19.8).

These results, consistent with previous psychological research (*see* Theoretical background), indicate that:

- The construction of children's mental maps of the home-school area relies at least partially on large-scale spatial skills (learning *environmental space* by locomotion and integration of information over time);
- The internalization of the configuration of larger spaces (test 2 and 3) that cannot be learned by experience but through symbolic representations e.g. maps, aerials or models (*geographical* → *figural space*) depends more on children's small-scale spatial abilities.

Therefore, primary school map education should recognize and support these both ways of learning spaces (especially when build up from local to global).

Further research could clarify if children's ability to construct a coherent geographical world view before leaving primary school could be reinforced through:

- Teaching methods that emphasize the conceptualization of the local environment starting from mental maps based on landmarks;
- Exploring spaces with digital cartographic tools, specifically designed for children, that enable adaptive zooming (large-scale, oblique view images with pictorial landmarks ↔ small-scale plan views with more abstract symbols).

## 19.5 Conclusion

Even by the end of primary school a majority of children in this sample still show a defragmented geographical world view. The lack of correlation between their representation of the familiar area versus larger spaces support previous research pointing out at least a partial dissociation between large-scale spatial abilities, needed for way-finding in the local neighbourhood, and small-scale spatial abilities, applicable when internalizing a large geographical space through symbolic representations such as maps and models. Further research could point out which teaching activities respond best to these findings and what new possibilities digital cartography has to offer.

**Acknowledgements** I am very grateful to the teachers and pupils for their willingness to participate in this study and I would especially like to thank student teachers Caroline Maes, Sofie Rosiers, Virginie Vandevelde and Liesbet Van Rosendaal who assisted me in the data collection. This research would not have been possible without their help.

## References

Catling SJ (1979) Maps and cognitive maps: the young child's perception. Geography 64:288–295
Downs R, Stea D (1977) Maps in minds. Harper & Row, New York
Harwood D, Rawlings K (2001) Assessing young children's freehand sketch maps of the World. Int Res Geogr Environ Educ 10:20–45

Hegarty M, Montello DR, Richardson AE, Ishikawa T, Lovelace K (2006) Spatial abilities at different scales: individual differences in aptitude-test performance and spatial-layout learning. Intelligence 34:151–176

Liben LS (2006) Education for spatial thinking. In: Damon W, Lerner R, Renninger KA, Sigel IE (eds) Handbook of child psychology, vol 4, 6th edn, Child psychology in practice. Wiley, Hoboken

Mark DM, Freska C, Hirtle SC, Lloyd R, Tversky B (1999) Cognitive models of geographical space. Int J Geogr Inf Sci 13:747–774

Matthews MH (1992) Making sense of place. Harvester Wheatsheaf, Hemel Hempstead

Montello DR (1993) Scale and multiple psychologies of space. In: Frank AU, Campari I (eds) Spatial information theory: a theoretical basis for GIS. Springer, Berlin

O'Keefe J, Nadel L (1978) The hippocampus as a cognitive map. Clarendon, Oxford

Panagiotaki G, Nobes N, Banerjee N (2006) Children's representations of the Earth: a methodological comparison. Br J Dev Psychol 24:353–372

Piaget J, Inhelder B (1956) The child's conception of space. Norton, New York

Quaiser-Pohl C, Lehmann W, Eid M (2004) The relationship between spatial abilities and representations of large-scale space in children—a structural equation modelling analysis. Personal Individ Differ 36:95–105

Roald I, Mikalsen Ø (2001) Configuration and dynamics of the Earth-Sun-Moon system: an investigation into conceptions of deaf and hearing pupils. Int J Sci Educ 23:423–440

Samarapungavan A, Vosniadou S, Brewer WF (1996) Mental models of the Earth, Sun, and Moon: Indian children's cosmologies. Cogn Dev 11:491–521

Schmeinck D (2007) Wie Kinder die Welt Sehen. Eine empirische Ländervergeleichstudie zur räumlichen Vorstellung von Grundschulkindern. Klinkhardt, Bad Heilbrunn

Scoffham S (1999) Young children's perceptions of the world. In: David T (ed) Teaching young children. Paul Chapman, London

Siegel AW, White SH (1975) The development of spatial representations of large-scale environment. In: Reese HW (ed) Advances in child development and behavior. Academic Press, New York

Siegel AW, Kirasic KC, Kail RV (1978) Stalking the elusive cognitive map. In: Altman I, Wohlwill JF (eds) Children and the environment. Plenum, New York

Spicer B (1984) Methods of measuring students' images of people and places. In: Haubrich H (ed) Perception of people and places through media: paper collection of the symposium of the commission "Geographical Education"; 25th international congress, Paedagogische Hochschule, Freiburg

Thommen E, Avelar S, Sapin V, Perrenoud S, Malatesta D (2010) Mapping the journey from home to school: a study on children's representation of space. Int Res Geogr Environ Educ 19:191–205

Wiegand P (1995) Young children's freehand sketch maps of the world. Int Res Geogr Environ Educ 4:19–28

Wiegand P, Stiell B (1996) Lost continents? Children's understanding of the location and orientation of the Earth's land masses. Educ Stud 22:381–392

# Chapter 20
# The Spatial Notions of the Cultural Universe of Childhood

**Paula Cristiane Strina Juliasz and Rosangela Doin de Almeida**

**Abstract** In Brazil, the area of school cartography counts on an event called Colloquium of Cartography for Children and Scholars. In the last edition of the Colloquium in 2009 themes which deserve some research were indicated, among them are the studies about 'cartography and childhood'. We elaborated this project with the objective to contribute with these studies. We are here investigating the elementary relations established by children aged 3–4 years old in the time-space-body organization. We will carry out an inductive analysis of the data registered during teaching activities which mobilized the time-space relations. This study has the objective to present the utilization of children literature as a way to understand these relations in the children's cultural universe.

## 20.1 Introduction

In school activities students and teachers re-elaborate their personal experiences, constructing and re-constructing knowledge, making the view from the inside of the school an essential element for the studies about teaching. Thus, the studies carried out in the Geography and Cartography Teaching Research Laboratory (LABENCARTOGEO—São Paulo State University, UNESP, Campus Rio Claro) take the school culture as reference. It is understood that the knowledge and school practices consist of social constructions, less connected to the knowledge prescribed in the curriculums, therefore founded in the knowledge and practice endowed with cultural values and with the school institution itself, which also fundaments the present study.

P.C.S. Juliasz (✉) • R.D. de Almeida
Geography and Cartography Teaching Research Laboratory, São Paulo State University, Bela Vista, Rio Claro, SP, Brazil
e-mail: paulacsj@rc.unesp.br; rda.doin@gmail.com

L. Zentai and J.R. Nunez (eds.), *Maps for the Future*, Lecture Notes in Geoinformation and Cartography 5,
DOI 10.1007/978-3-642-19522-8_20, © Springer-Verlag Berlin Heidelberg 2012

Studies developed in LABENCARTOGEO are the result of a trajectory of the Brazilian studies about Cartography teaching at school, initiated after the I Colloquium of Cartography for Children, held in Rio Claro—SP, in 1995. The event was held in 2009 and there some topics which need to be researched and studied were recommended, among them are the topics concerning childhood. Although there are several studies carried out by Brazilian researches about the genesis of the spatial representation in children's drawings, these studies do not reach kindergarten.

In order to meet this demand, we outlined this master's degree research entitled "A study about the cartographic language and the time-space representation by 4–5 year-old children" (this research is funded from The State of São Paulo Research Foundation, FAPESP). This research is mainly justified by the comprehension of time-space-body organization established by children and their representations, aiming to contribute to a childhood-related cartography. For this, we need to comprehend the motor-sensorial development of the child.

## 20.2 Time-Space-Body Organizations

With the comprehension of the fact that time and space are inextricably linked comes a pertinent question: how this relation is present in childhood, once space and time are fragmented and little by little acquire linearity and a sequential sense? One of the clues to answer this question is to consider that it is through the relations, which are established in the social groups in which they are inserted that the preschool children become aware of the different dimensions and relative values concerning time-space.

In this sense, we have an important reference on Vygostky and his successors for the analysis of the registers of teaching situations. Vygotsky's postulations about the biological and social factors in the psychological development point to two complementary ways of investigation: on one side, the knowledge of the brain as a material substrate of the psychological activity, and, on the other side, the culture as an essential part of the constitution of the human being, in a process in which the biological becomes social-historical (Oliveira 1992).

Considering this, the same author shows that the emergence of the orality allows a new structural organization of the action, attributing to the symbolic activity initiated with the speech an organizational function which produces fundamentally new forms of behavior (Vygotsky 2008). There is a convergence between the speech and the practical activity (action) in a way that the children control the environment through the use of the speech, before controlling their own behavior. There is a relation among time, space and speech. With the help of the speech, the children reorganize the visual-spatial field, evoking absent objects through the word and creating a temporal field that is perceptual, real and visual as well (Almeida 1994).

In the same line, Wardsworth (1989) presents the language as a social knowledge of adaptive value, that in the period between ages 2–4 or 5 is not characterized by the intention of communicating, and the thought that involves this language (a form of representation of objects and events) is free from the limitations of the direct action of the sensorial-motor thinking. With this, we can notice the relation between thinking and language. It is possible to observe the interaction of the psycho-physiological and social-cultural factors in the domain of the space, as many situations can be favorable for the development of particular competences, such as the ones that mobilize the *proprioception*, which refers to the sensibility of the movement of a particular part of the body or the whole body, or *kinesthesia,* differentiating the parts of the body.

Several studies based on the psychology of development and learning have indicated that the progressive acquisitions in the corporeal field amplify the domain of the space and that the posture influences the apprehension of the information about the environment. With this, special references are established in relation to the body itself, allowing the ingression in the founding factor of the special organization: the corporeal scheme.

According to the Wallonian perspective the corporeal scheme is the result of the relation that is established between the postural space and the environment space. The corporeal scheme is the cognitive basis on which the exploration of the space which depends both on the motor functions and the perception of the immediate space.

As soon as the children are aware of their oriented body, the temporal space increases and its geometry allows them to spread the axes of the body, serving as coordinates in the acquisition of the Euclidian space. When this happens, the articulation between the body and the more abstract and less limitative space is reached (Le Boulch 1982).

In this acquisition, it is possible again to notice the importance of the language in the formation of the concepts, making it possible to dissociate the corporeal scheme of the body itself and Project it on the objects. We project the orientations from our body (above/below, left/right, front/behind) in the space (Lurçat 1979). As for the language scope, (Le Boulch 1982) states that it allows the children to establish their references and relate them in a topological space, a process that can take place approximately at age 3.

The definition and orientation of the axes of the Euclidian abstract space influence the reintroduction of the body which will represent the true reference system (Le Boulch 1982).

In addition, the corporeal scheme is amplified through graphical representations. There are five stages in the construction of the human figure: the tadpole figure synthesis, its verticalization, the organization of the head-body structure, its improvement and the finishing with the drawing of well distinguished male and female figures (Greig 2004).

The corporeal experiences in the space influence the internal space organization, which are present in the representation of the human figure. These representations are originate in the acquisition of the closed forms that generate the tadpole figure in the representation of a character, which will later create an dissatisfaction in relation to its property to represent the body. Thus, concerning what we can consider a childhood-related cartography, the verticality of the body plays a structuring role. We think that this point of construction of the corporeal scheme is the fertile ground in which the notions about other spatial coordinates germinate. The verticality becomes the main axis of the whole human spatial organization (...). Therefore, it is in the childhood that the notion of *spatial coordinates* is originated. The drawing of a character is not only a drawing, for it brings in itself the germ of cartography (Almeida 2009).

Concerning verticality, while addressing the corporeal scheme and spatial organization, the vertical and horizontal references fundamental to constitute a system of coordinates which establish relations of order which will in turn orient the Euclidian space, after the topological space (Le Boulch 1982).

Other authors have addressed the representation of the space, among them and of great importance is Jean Piaget, author of *Space representation in the child*, co-written by Barbel Inhelder (Piaget and Inhelder 1993). It is important to consider that Piaget's studies did not consider the space as a geographic concept. The authors' concerns were related to the mathematical and geometrical space, which although not referring to the terrestrial space the same way as geography, constitutes the basis of Cartography (Almeida 2001). The theory that Jean Piaget constructed with the help of a group of researchers remains as a fundamental theoretical support for the studies about the representation of the space, mainly because it addresses the construction of the mathematical space by the children (topological, projective and Euclidian relations) on which the geographical space leans (Almeida 1994).

The sensorial-motor activities and their relation with the corporeal scheme are fundamental in the construction of the space by the child. Therefore, we developed a number of teaching activities that mobilize the time-space-body organization with the objective to analyze the performance of children aged 3–4 in learning situations.

## 20.3 The Investigation Path

To investigate the relations that children establish in the time-space-body organization we will carry out an analysis of the data obtained from ethnographic registers, based on the methodology of the qualitative research (Bogdan and Biklen 1994).

In this approach, the presence of the investigator in the place of the studies allows the understanding of the actions of the individuals who are to be studied, their activities and environment, aiming at the approximation between researchers and researched. The qualitative method has as basis the intensive and long-term participation in the field, accurate registers about what happens in the field, analytic reflection on these data and written register of the participative observations (Erickson 1989). Under this perspective, the researcher should try to find ways to understand the meaning of the manifest and latent of the behaviors of the individuals, while trying to maintain an objective view of the phenomenon.

The researcher plays the subjective role of participant and the objective role of observer, reflecting about the experiences of the students and about the narratives produced by these observations and recordings of activities in the classroom. This type of narrative investigation is a variable of the qualitative research, being justified by the fact that we—human beings—are story-tellers organisms that, individually and socially live reportable lives (Connelly and Clandinin 1995).

After the activities performed by the children, narratives about the facts and students' dialogues are prepared, allowing the emergence of several "selves" through the interventions and students' productions, the reflections on the activities and aspects of the process (Conelly and Clandinin 1995). In this perspective, both the teacher and the students are story-tellers and also characters of their own and other stories. Therefore, the narrative structures the experience that will be studied and will be the way by which it will be analyzed and how it will be presented.

As an example, we will present a teaching activity that was performed with children from kindergarten. It is important to emphasize that the dialogues exposed here use the initial letters of the students' names and TEACHER refers to the researcher who performed the practical activity.

### *20.3.1 Cartography in the Children's Universe*

To perform this activity, we chose a children's literature book because this kind of book mobilizes the imaginary of the children and is widely used by the teachers. We used the book "A Pirilampéia e os dois meninos de Tatipurum" (The Pirilampéia and the two boys in Tatipurum), by José Rufino dos Santos (Santos 2000), a story of two boys who live in Tatipurum (a planet). Each boy lives in one side of the planet so they always argue about who is upside down.

A cicada called Pirilampéia, helps them to find out that in the space there is no upside or downside, no matter the side of the planet you are on. From this story we notice that the spatial relations of neighborhood (upside/downside) could be developed in the classes.

We planned a didactic sequence composed by three activities developed in a class (Pré I A) with kids aged from 3 to 4 years old, at "Escola Municipal Sueli Maria Proni Cerri", in the city of Rio Claro, Sao Paulo. In this study we will present the first stage of this didactic sequence, which has the purpose to get to know and represent the story in a tridimensional material.

The activity was recorded—image and sound—making it possible to transcribe the speech of the participants and also the gestures of the students, elements that sometimes escape from the teacher-researcher's immediate perception. For this transcription, we used the minuting technique, which consists in taking notes of the observation in intervals, and when necessary integrally transcribe parts of the participants' dialogues. These transcriptions helped in the elaboration of the narrative that reports the development of the activity.

To perform the activity, we divided the students in two groups so that we could register the students' dialogues and actions more accurately. ANA, GAB, JES, JHE,

MAR, VIN, participated in the activity, and they were chosen using affinity as criteria, because the activity involved pair-work.

We had the students sit around a table and started the story by the observation of the book cover and reading the title "A Pirilampéia e os dois meninos de Tatipurum" (Pirilampéia and the two boys from Tatipurum). In the beginning, the characters "Tom" and "Dick" are introduced, and to make it easier to distinguish them we talked about the color of their hats, Tom wearing a blue hat and Dick a green one. JES anxiously asked what Tatipurum was. We showed that it was a planet in which the boys lived.

(TEACHER) Tom and Dick lived in planet Tom lived in one side, Dick in the other—each character is shown on different pages of the book.

So we moved on, reading the story using the book's illustrations:

(TEACHER) Tom and Dick did not have much to do. Tatipurum was a boring planet. Tom was always chasing ants. He put them on the palm of his hands and blew the little creatures to space. Dick tried to spit farther and farther. Each time he tried to break his own record. One day, tired of this game, he shouted to Tom: 'Hey, boy! Do you like to stay upside down?'

While reading this excerpt, some children laughed and looked curiously at the illustrations. And, in this moment, the question "who is upside down" appears as the central conflict of the story, which promoted some questioning from students like GAB, who asked:

(GAB) I want to see him upside down! Where is it?
(TEACHER) He is saying that he is right side up and that the other is upside down. Let's see who is upside down—I answered—So, the other said: it is you who is upside down. I am right side up.

To prove that he is not upside down, Tom plants a tree called Jameleira and says "Jameleira grows upwards, doesn't it? So, if I am upside down it doesn't grow". When they saw the illustration of the planet Tatipurum with this tree the students asked to turn the book so that the illustration of the Jameleira was vertical.

Continuing the story, Dick made a balloon and asked Tom: "a balloon goes to what direction?". Tom said that the balloon goes upwards, so Dick stated that if he were upside down the balloon would not go upwards. The balloon went upwards and the two boys began to fight. Then appears a cicada called Pirilampéia, in the planet Tatipurum, She came from the planet Pirilampeu. She wanted to know why the boys were fighting and when she understood what was going on she explained: "I, who come from space, can land on Tom's space. Then, Dick is upside down." Then, Pirilampéia flew to Dick's side and said: "Now it is Tom who is upside down". And I finally said that in the space there is no upside or downside, asking the students to change sides.

To talk about this change I said "one here" and gesticulated. The student JHE also gesticulated and said "one there". These gestures show the notions of near and far related to here and there.

**Fig. 20.1** Material used in the activity: Tatipurum, Pirilampéia, Tom, Dick and Jameleira

When we finished telling the story, we asked the students to get in pairs. I informed that we would perform an activity with the materials I had brought (Fig. 20.1). These materials consisted of a green foam ball and the pictures of the following elements: Pirilampéia, Tom, Dick, Jameleira and the balloon. The students were supposed to place them in the correct place on the foam ball.

We distributed the material to each pair which caused some conflicts among the students, as in this stage of development egocentrism is a characteristic of the children. Some students explored the material rolling the foam ball on the floor or throwing it up. After, we gave them character Tom.

(TEACHER) Who is this?
(JES) Tatipurum.
(TEACHER) Tatipurum is the planet that is on your table. This boy wearing a blue hat is Tom.

Then we gave them character Dick. We asked them to place the two characters (Tom and Dick) in the planet "Tatipurum". We observed that some students as JHE had already started to do this, before we asked them to do so. However, we noticed that the pair MAR and JES was not doing what they were supposed to, instead they were disputing the material. JES asked me which figures were MAR's, and it was necessary to talk about the division of the material and the proposal of the pair work.

As the pairs placed the characters, they talked a little about their actions. JES stated that the characters were organized that way because MAR "did it the wrong way":

(TEACHER) Did it the wrong way? What do you mean?
(JES) She was supposed to put hers here and not where mine is.
(TEACHER) And, why do you think that hers is supposed to be put there?—trying to understand if what she was saying had any relation with the organization of the characters in the story.
(JES) Because it cannot be put beside mine
(TEACHER) Why don't you want one beside the other? (JES) Because not.

Trying to understand MAR's position in the activity, we asked:

(TEACHER) MAR, why did you want to put it in this side?
(MAR) Because I wanted to do so.
(JES) But I didn't!

We observed that this pair was more concerned about the object itself instead of considering the story and there was evident dispute between the children despite their affinity. Answering "because I wanted to do so" can be related to the egocentrism t can also indicate that one of the students found out that the characters occupied opposite positions and the other did not see this clearly.

The pair ANA and GAB stated that they had placed the characters one in each side 'because we like it this way'. The students shared the material in the following way: GAB stayed with Dick and ANA with Tom. To show the way they had placed the characters ANA placed Dick in the opposite side. In this case, we noticed that the character placed by GAB served as reference to ANA. GAB, talking about the position of the characters concluded: 'this here and that there'.

In the third pair, JHE stated that the characters were one in each side in the activity because they were in this position in the book. Then, with the characters placed on Tatipurum, I introduced other elements of the story that would also be placed on the foam ball: Pirilampéia, Jameleira and Ballon. Some questions emerged:

(JES) Which of them made the balloon? Let's see in the book? Does anyone know who made it?
(JHE) The green one.
(TEACHER) Right. The one with the green hat, Dick, made the balloons. The one with the blue hat, Tom, planted the Jameleira.

We felt the need of a referential which would allow the exploration of the notion near/far, using the objects created by the characters and their location on the planet Tatipurum. We observed that JHE and VIN were disputing the material and were not in agreement about the position of the balloon. VIN indicated that he would place it beside Tom, but before he had said that Dick had made the balloon and JHE stated that it should be placed nest to Dick. VIN's suggestion prevailed. The children's productions can be seen in Figs. 20.2, 20.3 and 20.4.

To finish the activity, we had the students sat together and started a conversation:

(TEACHER) The balloon near Dick and Jameleira near Tom, because he planted it. And Pirilampéia?—observing the material used by ANA and GAB.

**Fig. 20.2** Production by JES and MAR, elements of the story focused on one part of the surface of the foam ball

**Fig. 20.3** Production by ANA and GAB with position of the elements of the story in the foam ball

**Fig. 20.4** Production by JHE and VIN with position of the elements of the story in the foam ball

(GAB) To help them.
(TEACHER) What kind of help?
(GAB) Prevent them from fighting
(TEACHER) What did she say to make them stop fighting?
(JHE) To change.
(TEACHER) Let's change here, in your activity?

The children accepted the suggestion and so we changed the position of the characters. We observed that JHE changed the position of the characters and other elements as well, placing the balloon close to Dick and Jameleira close to Tom, actions that VIN did not adopt despite his suggestions. This demonstrates the importance in developing individual activities with children from this age group when the objective is to understand individual notions related to topological notions. To help in the comprehension we asked:

(TEACHER) What did they notice when they changed places?—There was silence and I moved on—They noticed that there was no upside or downside, as Pirilampéia said: "when I am here, Tom is upside down and if I fly here it is Dick who is going to be upside down. But in space there is no upside or downside"—I said using the concrete material.

This finalization—using the specific material to explore the problem—was positive, as the students were able to interact and explore once more the narrative.

## 20.4 Final Considerations

Children's literature is really a bridge to the universe of childhood. We observed that as the story advances a sequence of facts is established which pre-establishes the temporal sequence and that the reading of stories can promote the construction of linearity in the relation time-space.

In this activity we could observe that the students took the story as reference to locate certain elements, as in the case of the tree and the balloon (JES asked the peers if they knew who had made the balloon before placing it). Some children used a previous action as reference, as ANA, who placed the character after GAB had done it. We can state that some children may have placed their characters according to the illustrations of the book (JHE anticipated the explanation of the activity placing one of the boys in a side of the planet).

After this activity we noticed that it was important to perform more activities to promote the relations of neighborhood and that it would be necessary to carry out an individual activity to verify the performance of each student. So, two other activities were planned. The first consisted in the organization of the elements of the story through pasting (using bidimensional material) and the second a drawing about the story.

The students' productions are "registers" by the children themselves concerning the theme and can be filed in an individual portfolio. In addition to these materials, the dialogues show that this kind of activity promotes a collective discussion about some topics allowing the production of knowledge by the interaction among people.

We emphasize the importance of the techniques used to register the activities, because they enabled us to obtain accurate information about the students' thinking (in the dialogues transcribed in the recording minuting) As a result of this research we hope to obtain a rich collection of registers and narratives to be analyzed in the future, and also clarify some aspects about the relations time-space-body and their representation.

## References

Almeida RD (1994) Uma proposta metodológica para a compreensão de mapas geográficos. Dissertation, Universidade de São Paulo, Faculdade de Educação, São Paulo

Almeida RD (2001) Do desenho ao mapa. Contexto, São Paulo

Almeida RD (2009) Cartografia e Infância. VI Colóquio de Cartografia para Crianças e II Fórum Latino-americano de Cartografia para Escolares. Juiz de Fora—MG. http://www.scribd.com/doc/21198272/Cartografiaeinfancia?secret_password=1y7uc0bv6objyzc80osp/. Accessed 10 Feb 2010

Bogdan R, Biklen S (1994) Investigação Qualitativa em Educação: uma introdução à teoria e aos métodos (trans: Alvarez MJ, dos Santos SB, Baptista TM). Porto Editora, Porto

Connelly FM, Clandinin DJ (1995) Relatos de Experiencia e Investigación Narrativa. In: Larrosa J et al (eds) Déjame que te Cuente: ensayos sobre narrativa y educación. Alertes, Barcelona, pp 11–59

Erickson F (1989) Métodos cualitativos de investigación sobre la enseñanza. In: Wittrock MC (ed) La investigación de la enseñanza II. Métodos cualitativos y de observación. Paidós, Barcelona, pp 195–299

Greig P (2004) A criança e seu desenho: o nascimento da arte e da escrita. Artmed, Porto Alegre

Le Boulch J (1982) O desenvolvimento psicomotor: do nascimento até 6 anos (trans; Brizolara AG). Artes Médicas, Porto Alegre

Lurçat L (1979) El niño y el Pespacio; la función del cuerpo (trans: Zenzes EC). Fondo de Cultura Económica, México

Oliveira MK (1992) Vygotsky e o Processo de Formação de Conceitos. In: La Taille Y, de Oliveira MK, Dantas H (eds) Piaget, Vygostky, Wallon: teorias psicogenéticas em discussão. Summus, São Paulo, pp 23–34

Piaget J, Inhelder B (1993) A representação do espaço na criança (trans: de Albuquerque BM). Artes Médicas, Porto Alegre

Santos JR (2000) A Pirilampéia e os dois menino de Tatipurum. Ática, São Paulo

Vygotsky LS (2008) Pensamento e linguagem (trans: Camargo JL). Martins Fontes, São Paulo

Wardsworth BJ (1989) Inteligência e afetividade da criança na teoria de Piaget. Livraria Pioneira, São Paulo

# Chapter 21
# Map Drawing Competition for Children in Indonesia

**Rizka Windiastuti**

**Abstract** The National Coordinating Agency for Surveys and Mapping of Indonesia (BAKOSURTANAL) has been organizing map drawing competitions for children in Indonesia as a way to popularize map to children. These competitions were always handled seriously and such competitions were always participated by more than expected numbers of children. This paper will share how BAKOSURTANAL organized the competitions, from the publication until the judgments to get the winners.

## 21.1 Introduction

The National Coordinating Agency for Surveys and Mapping of Indonesia (BAKOSURTANAL) is a government body that is responsible to conduct governmental duties in the field of surveys and mapping according to prevailing regulations. As the coordinating agency for survey and mapping activities in national scope, BAKOSURTANAL has an obligation to disseminate information about services and products of survey and mapping activities, especially the ones performed by BAKOSURTANAL. It is also necessary to raise awareness about maps and the usages to various levels of community, including the children. Some of the methods used to raise public awareness include organizing various exhibitions, workshops, seminars, trainings, map adventure games, and map drawing competitions.

The map drawing competition was initially inspired when the International Cartographic Association (ICA) in cooperation with the United Nations Children's Fund (UNICEF) invited BAKOSURTANAL to coordinate a map drawing competition

R. Windiastuti (✉)
Geomatics Research Institute, National Coordinating Agency for Surveys and Mapping of Indonesia (BAKOSURTANAL), Cibinong, Indonesia
e-mail: rizka.windiastuti@gmail.com

L. Zentai and J.R. Nunez (eds.), *Maps for the Future*,
Lecture Notes in Geoinformation and Cartography 5,
DOI 10.1007/978-3-642-19522-8_21, 

in Indonesia. The best five drawings were then sent to Durban, South Africa, to be enrolled in an international map drawing competition held during ICA Congress in 2003. That was the first time BAKOSURTANAL, in cooperation with ICA and also Indonesian Cartographic Association (AKI), organized a national map drawing competition.

Following the success in 2003 and considering the number of children participated in the competition, BAKOSURTANAL organized the next national map drawing competitions in 2005, 2007, 2009 and 2011, and always sent the best drawings to go into Barbara Petchenik competitions held by the ICA in the year. In addition to the national competitions, BAKOSURTANAL also held local competitions in various areas in Indonesia:

- 2005 in Cibinong (West Java),
- 2006 in Surabaya (East Java) and Jakarta,
- 2007 in Jakarta,
- 2008 in Bandarlampung (Lampung),
- 2010 in Pontianak (West Kalimantan), (BAKOSURTANAL, 2010)

Children map drawing competitions, which allow the children to explore their imaginations in drawing, have been used by BAKOSURTANAL to introduce maps to children. By following such competition, children will be driven to open their atlases and learn about the world, such as the richness and beauty of their country or even province, through maps. Hopefully, this competition would (directly or indirectly) contribute to Cartography education in Indonesia, and to increase the children's care and respect to their country and the Mother Earth.

## 21.2 Organizing Map Drawing Competitions

The map drawing competitions held by BAKOSURTANAL always participated by an unexpected number of participants. The national competition in 2003 was followed by 1,045 participants from 27 provinces in Indonesia, in 2005 by 1,002, in 2007 by 1,377, in 2009 by 1,179, and in 2011 by 1,205 participants. The local competitions in various regions in Indonesia were also interested by no less than 200 children. Here will be explained how BAKOSURTANAL organized the competitions.

### *21.2.1 Preparation*

The preparation to organize a map drawing competition was started since about the middle of the preceding year, i.e. when BAKOSURTANAL prepared the budget for activities in the subsequent year. The organization of a national map drawing competition was somewhat different from that of a local competition. For the national competitions children drew at home or school then sent the result to

BAKOSURTANAL, so the children had plenty of time to draw. For the local competitions the children came to the venue and drew on the site for about 3 h.

Initial stage in executing a map drawing competition was to form an organizing committee at the beginning of the year when it will be run. The organizing committee of the national competitions comprised about 10–15 staff of BAKOSURTANAL, where as for the local competitions, especially the ones held outside Jakarta or Cibinong area, BAKOSURTANAL collaborated with other institutions as the local hosts. In 2006 we cooperated with Surabaya State University (UNESA), in 2008 with Yayasan Xaverius Tanjungkarang, and in 2010 with Tanjungpura University—Pontianak.

In general, the organizing committee consisted of three teams:

- Secretariat team who was responsible to do administrative tasks, accept the drawings (for national competitions) or registrations (for local competitions), collect data about the drawings, and answer any questions regarding the competitions that were asked by the public.
- Publication team who was responsible to let general public know about the ongoing competition through advertisements in mass media (national newspaper, radio, etc.) and/or brochures that were distributed to schools and other public areas.
- Selection team who was responsible to communicate with the juries about selection criteria, prepare and assist the juries during the judgment.

In addition to the organizing committee, at this initial stage were also formed boards of juries. For the national competitions, because we expected to have more participants than the local ones, we prepared two kinds of juries, namely: internal juries and national juries. The detailed duties of these juries will be explained later in this paper. The juries were selected considering their knowledge in cartography and art.

The organizing committee would then create a theme and rules for the competition and set up some important dates, including: when were the promotions, the last day to receive drawings/registration, the judgment, the winner announcement, and the delivery of best drawings to ICA (for national competition). The theme and rules for national competitions would be the same as those of Barbara Petchenik competitions held by ICA, including the drawing groups based on the ages of the authors (group A for ages below 9 years old, group B for ages 9–12, and group C for ages 13–15). For local competitions, we classified children into three groups based on their school grades (group A for grade 1–3, group B for grade 4–6, and group C for grade 7–9).

### 21.2.2 Drawing Collection

After the organizing committee was formed and the information about the competition was open to public, the secretariat team must be ready to accept registration or

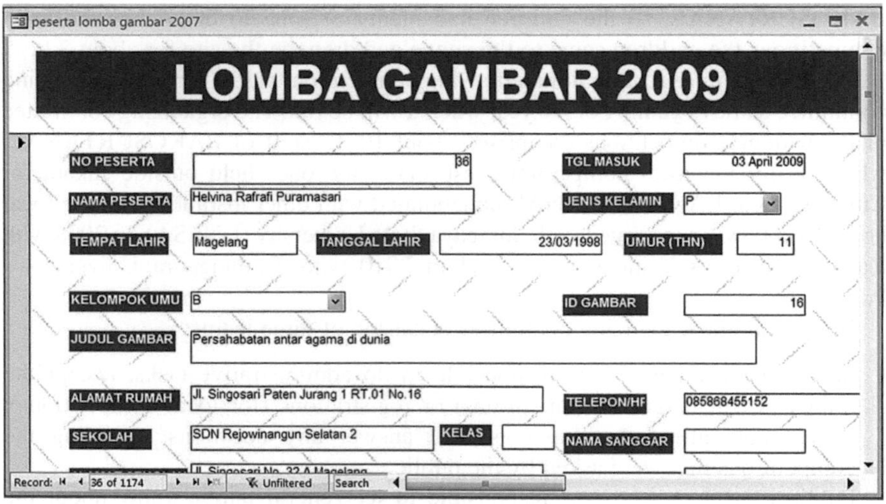


**Fig. 21.1** Database form to enter participant's data during national competition in 2009

to collect the drawings by preparing a database to store data of the participants. The database was developed using Microsoft Access and for the national competitions it included data about the authors (name, school and home addresses, date of birth, etc.) and the drawings (title, received date, etc.) (Fig. 21.1). It was important to record as many details as possible about the participants' addresses and phone numbers because later we would need to get in touch with some of the authors. For local competitions the database prepared for the registrants was much simpler. Every drawing received for a national competition was stamped and given a unique ID number. Data entry for the database was performed soon after the drawings arrived.

The data entry was a crucial step during execution of map drawing competitions (Fig. 21.2). We used the data entered in previous years to inform participants about succeeding events. For local competitions, the registrant data were used on the competition day, when children came to the venue and they were asked to re-register so they could get an ID badge, T-shirt and snacks. Children who did not register beforehand were allowed to register at the time as long as there was space available for them.

During the national competitions, where the participants sent drawings to our address, the packages were usually delivered near the end of the deadline. Figure 21.3 below showed the number of drawings received on a specific day during the national competition in 2007. The deadline at the time was post stamped on 30 March 2007. We can see from the graphic that most participants sent the drawings on or slightly before 30 March 2007 such that most drawings arrived on 2 April 2007 (BAKOSURTANAL, 2007). From this experience, since the following national competition in 2009 the deadline was no longer the delivery date, but the date the drawings were received by the organizing committee.

**Fig. 21.2** Organizing committee entering drawing data into the database

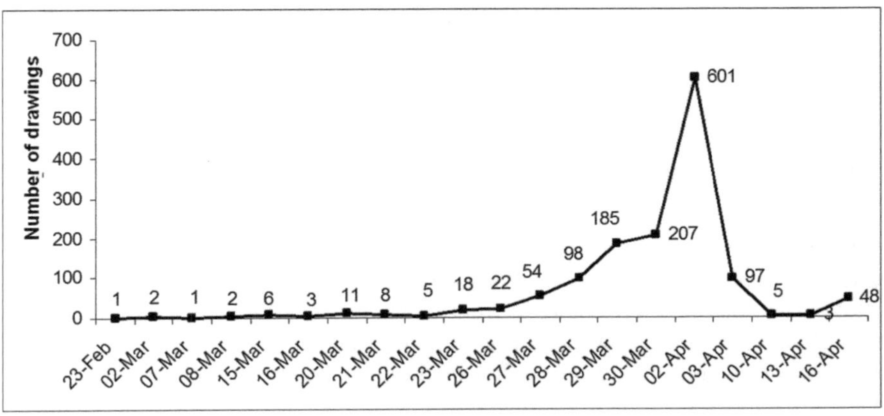


**Fig. 21.3** Number of drawings received on a specific day during national competition in 2007

## 21.2.3 *Judgment*

### 21.2.3.1 National Competition

As mentioned previously, for national competitions we had two boards of juries: internal and national. The internal juries comprised of three to five staff of BAKOSURTANAL who met the criteria as a jury for map drawing competitions. The national juries comprised of three persons; at least one of them came from BAKOSURTANAL with a strong knowledge about cartography, and at least one of them came from an art institute or was a professional artist who had experience in judging children's drawing competitions.

The mechanism of judgment for national competitions was as follow:

- Before being evaluated by the boards of judges, the drawings were selected by the selection team of organizing committee based on their administrative completeness. The drawings that used paper size other than A3, had incomplete author's identity (such as no address or date of birth written either on the back of the drawing or on a separate piece of paper) and were out of the predefined theme, were eliminated. In fact, many drawings did not include maps or only had maps, so these drawings were considered out of theme.
- If the number of drawings were more than 250, then the judgment would be done in three steps for every group; the first step was performed by the internal juries to eliminate about a third or half of the drawings, the second step was performed by the national juries to select 15 best drawings for each group, the last step was also performed by the national juries to select the best drawings to be enrolled into Barbara Petchenik competition and additional three best drawings from each group.
- If the number of drawings were equal to or less than 250, then the judgment would be done directly by the board of national juries to select 10–15 best drawings for each group, then select the best drawings to be enrolled into Barbara Petchenik competition and additional three best drawings from each group.

During the selection process performed by board of internal juries, each of the juries was given an evaluation sheet to evaluate the drawings, as shown in Fig. 21.4 below. The drawings were hung in front of the juries, 20 drawings at a time, and each of the juries decided whether each drawing had the following criteria: composition of world map, scale proportion of the map features (applicable for group B and C), relative position of the map features (applicable for group C only), theme suitability, color composition, and creativity. For every drawing, each jury put a check mark on the criteria that existed in the drawing, and let the organizing committee do the counting.

During the selection process by board of national juries, it is important for the juries to be able to see all drawings at once (Fig. 21.5). Therefore, all drawings from each group were put on the floor with the drawing side up. The juries then looked

**LEMBAR PENILAIAN TAHAP I**
**LOMBA GAMBAR PETA UNTUK ANAK**
**BAKOSURTANAL 2009**
**KELOMPOK A**

*Penilai :*

| No | No. Gambar | Kriteria | | | | | | |
|---|---|---|---|---|---|---|---|---|
| | | Komposisi peta dunia | Proporsional skala | Posisi relatif peta | Kesesuaian tema | Komposisi warna | Kreativitas | Jumlah |
| 1 | | | | | | | | 0 |
| 2 | | | | | | | | 0 |
| 3 | | | | | | | | 0 |
| 4 | | | | | | | | 0 |
| 5 | | | | | | | | 0 |

**Fig. 21.4** Evaluation sheet for internal juries

**Fig. 21.5** Judgment by board of internal juries

around and eliminated the drawings by turning the drawing side down. It was not so hard in the beginning, but as the drawings got fewer, the juries would discuss to determine the best 15 drawings (Fig. 21.6).

Figure 21.7 showed the result of judgment during the 2009 competition. At that time we received 1,179 drawings, 13 were sent by children whose ages above 15 years old and 5 had no author's identity, thus 18 drawings were eliminated from the competition at the beginning.

After we found the 45 finalists, the juries would then determine the overall winners who would represent Indonesia in Barbara Petchenik competition.

**Fig. 21.6** Judgment by board of national juries

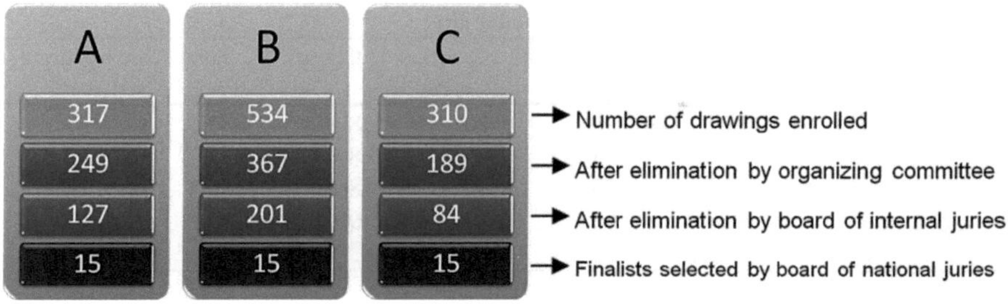


**Fig. 21.7** Elimination process during judgment of national competition in 2009

The juries individually nominated 2 drawings for each group, totaling 18 drawings. A drawing that was nominated by most juries automatically became an overall winner. The juries would then discuss to select the drawings nominated by only one jury until we obtained five overall winners. Each jury would then give a score of 1–100 to the rest of 40 finalists, considering map contents, theme suitability and creativity. The score was necessary in case winning candidates failed verification process.

### 21.2.3.2 Local Competitions

Local competitions were usually held from 9 AM to 12 PM in a hall provided by the local hosts. The children, unaccompanied by their family members, sat on the floor according to their ID numbers. After all participants sat in the place assigned to them, organizing committee distributed drawing papers to be used. The judgment process begun soon after the organizing committee collected all the drawings.

Board of juries for local competition usually comprised of three persons: at least one of them came from BAKOSURTANAL and at least one jury taken from a local paint artist who had experience as a jury for children's drawing competitions. Similar to the judgment process of national competitions, here all the juries together determined the best 15 drawings from every group, to get the total of 45 finalists. The winners were determined by the total scores given by all juries.

During the scoring stage, every jury individually evaluated each drawing based on certain criteria, which are: theme suitability, map contents, creativity, color composition, and drawing technique. Each component of the criteria was given a weigh factor as follow:

- Theme suitability: 0.30,
- Map contents: 0.20,
- Creativity: 0.20,
- Color composition: 0.15,
- Drawing technique: 0.15.

**Table 21.1** Sample of initial scoring by individual jury

| Drawing ID | Criteria #1 | Criteria #2 | Criteria #3 | Criteria #4 | Criteria #5 | Total score by jury #1 |
|---|---|---|---|---|---|---|
| 1 | 95 | 90 | 80 | 80 | 80 | 86.5 |
| 2 | 90 | 95 | 85 | 80 | 80 | 87 |
| 3 | 95 | 90 | 80 | 80 | 80 | 86.5 |
| **4** | **95** | **95** | **90** | **90** | **90** | **92.5** |
| ... | ... | ... | ... | ... | ... | ... |

**Table 21.2** Sample of final scoring sheet

| Drawing ID | Participant ID | Score by jury #1 | Score by jury #2 | Score by jury #3 | Total score |
|---|---|---|---|---|---|
| 9 | 26 | 93.65 | 76.85 | 93.95 | 264.45 |
| 8 | 21 | 94 | 73.95 | 95.7 | 263.65 |
| **4** | **20** | **92.5** | **75.85** | **94.95** | **263.3** |
| 11 | 2 | 79.25 | 75.95 | 92.1 | 247.3 |
| ... | ... | ... | ... | ... | ... |

The juries gave a score of 1–100 to every component for each of the 45 finalists. The organizing committee then assisted the juries to calculate the final scores. Score given by each jury for each drawing was the sum of the score of every component multiplied by its weigh factor. The final score for the drawing was the sum of its score by all juries. Tables 21.1 and 21.2 illustrate how the final score of the drawings were obtained.

## *21.2.4 Verification*

For the national competitions, since the children drew at home or school and sent the result to the organizing committee, it is important to verify that it was indeed the children who did the drawing. For that purpose, we established a verification process in which we visited the winning candidates at their homes or schools and asked the children to draw. There were two things that need to be verified during the visit:

- That the children really did the drawing by themselves from scratch until finished.
- That the children really belonged to an age group as informed to us, and thus were eligible to follow the competition (from the children's date of birth shown on their birth certificate or official school report).

During the verification process, it is important to make the children as comfortable and confident as possible (Fig. 21.8). While the children were drawing, we interviewed the parents or the teachers, and sometimes also the children and their friends, in order to get some information about the children's background. At this opportunity we also asked permissions from their parents to publish the winning drawings and enroll the drawings to the competition held by ICA (Fig. 21.9).

Fig. 21.8 Verification of winning candidates at her school

Fig. 21.9 The champion's drawing of group A (grade 1–3) at local competition in 2008, entitled "Natural Resources of Lampung", created by Syifa Putri Atalia Sadil in less than 3 h (BAKOSURTANAL 2008)

**Fig. 21.10** A drawing by Juan Edwin (10 years old) entitled "Living in a globalized world in one ark", which was enrolled into ICA competition in 2011

### *21.2.5 Winner Announcement*

Winners of local competitions were announced soon after the final scores were obtained, and the presents were given directly to the children. BAKOSURTANAL provided presents in forms of money, BAKOSURTANAL's products, trophy, certificate and school equipment (BAKOSURTANAL, 2005).

Winners of national competitions were announced at BAKOSURTANAL website, national newspaper, and a letter was sent to each winner (Fig. 21.10). The winners who resided around BAKOSURTANAL office were invited to pick up their presents. For those who lived in other areas, the presents were delivered by mail and the money was transferred to their bank accounts.

## 21.3 Conclusion and Recommendation

BAKOSURTANAL has been successfully organized map drawing competitions for children in Indonesia since 2003. The high number of participants showed that the information about these competitions was well distributed. However, most of participants for national competitions were from Java Island. For example, percentage of participants coming from six provinces in Java Island was 73.93% in 2007 and 83.88% in 2009. This data suggested that BAKOSURTANAL needs to socialize maps to children outside Java Island, perhaps through local map drawing competitions. BAKOSURTANAL can also establish partnership with regional institutions, such as universities, local government or even painting studios to organize map drawing competitions in the region. Hopefully, BAKOSURTANAL's task to raise awareness about maps and the usages to entire levels of community can be well implemented.

## References

BAKOSURTANAL's Website (2008) Lomba Gambar Peta Untuk Anak Se-Provinsi Lampung. http://www.bakosurtanal.go.id/bakosurtanal/lomba-gambar-peta-untuk-anak-se-provinsi-lampung/. Accessed 19 Mar 2011

BAKOSURTANAL's Website (2010) Lomba Gambar Peta Untuk Anak di Pontianak. http://www.bakosurtanal.go.id/bakosurtanal/lomba-gambar-peta-untuk-anak-di-pontianak/. Accessed 19 Mar 2011

Warta BAKOSURTANAL (2005) Lomba Gambar Se-Jabodetabek dan Tingkat Nasional 22 Mei 2005. http://www.bakosurtanal.go.id/bakosurtanal/warta-bakosurtanal-edisi-ii-juni-2/. Accessed 19 Mar 2011

Warta BAKOSURTANAL (2007) Indonesia Siap Kirim Wakilnya di Kompetisi the Barbara Petchenik Award. http://www.bakosurtanal.go.id/bakosurtanal/indonesia-siap-kirim-wakilnya-di-kompetisi-the-barbara-petchenik-award. Accessed 19 Mar 2011

# Chapter 22
# Cartography in Studying the Environment: Bilingual Practice Aiming at the Inclusion of Deaf Pupils

**Tiago Salge Araújo and Maria Isabel Castreghini de Freitas**

**Abstract** When it comes to Special Education, there are countless proposals, debates and disagreements. In Brazil, the fact is that special schools have been extinguished by yielding more place to heterogeneous schools that are full of diversity. Regarding deaf pupils, they must highlight their own language and culture; therefore, they are ultimately required to study and foremost to develop "a desire to integrate". This research was mainly aimed to identify possible methodologies and activities targeted to the deaf pupils entered into the regular Education. Because it is a qualitative research on a school located in the outskirts of a city in Brazil, the topic of urban environmental problems was chosen to develop the activities of the pupils Assuming that to intervene in the problems of the environment, at first, we must realize the place where we live in, Cartography is an important instrument used in the activities proposed by this project.

## 22.1 Introduction

In 2008 I was charged to replace a professor of Geography at the municipal schools in the region of Rio Claro, São Paulo, Brazil. In one of the classes I had two deaf pupils. The replacement period of 2 months was insufficient for a comprehensive study. However, I had experience with these pupils. The difficulties of orientation were combined with the infrastructure offered by the school to work with this particular class. The special needs of deaf pupils have clearly shown during the lessons that being included in the regular education system often does not ensure

---

T.S. Araújo (✉)
Faculdade de Psicologia e de Ciências da Educação, Universidade do Porto, Porto, Portugal
e-mail: salarau@hotmail.com

M.I.C. de Freitas
Universidade Estadual Paulista, UNESP, São Paulo, Brazil
e-mail: ifreitas@rc.unesp.br

L. Zentai and J.R. Nunez (eds.), *Maps for the Future*,
Lecture Notes in Geoinformation and Cartography 5,
DOI 10.1007/978-3-642-19522-8_22, 

linguistic equality and knowledge acquisition. It was this short experience that motivated me to investigate the subject and explore issues relating to this situation. This paper includes the proposals of this research.

The objective of this research is to plan and develop teaching activities focusing on geography, in particular with regard to spatial perception, which relates to the different methodological and linguistic needs of a mixed class of regular schools with deaf pupils.

For several centuries, the deaf have been struggling to articulate their goals of conquering space for participation, engagement and dialogue in our society. Just by understanding them as a cultural group that shares the history and linguistic situation, the paper we will adopt the terminology *deaf*. To cite Skliar (2001), sign language nullifies the deficiency and allows that the deaf are, then, a linguistic minority community different and not a deviation from normality. This proposal aims to share the experiences of working methodology with other professional teachers who were faced daily with deaf pupils in their classes. First, the practices related to this research would be directed to study and activities with maps, charts and aerial images, however, the mapping developed in this work was very far from activities with coordinates, scales and types of representations.

As the project was developed and as it unfolded, new activities have emerged and been proposed in an attempt to better understand and also to encourage deaf pupils in their perceptions of living space. However, with regard to the effectiveness of the process of school integration, we must emphasize that this necessarily depends on the cooperation of everyone involved: from the teachers to the pupils participating in the activities. As the qualitative research proposed, the information is not collected to confirm or disprove hypotheses constructed previously. The abstractions were built as private data were gathered (Bogdan and Bilken 1994). Based on Bogdan and Biklen (1994), our planning was done throughout the research and data analysis. The data have been analyzed more systematically in the final stages of research. It is also important that the bases for the planning of learning activities are well defined, as well the materials and teaching methods adopted, "interlacing the individual parts" (Bogdan and Biklen 1994). For being deaf children with their own language and culture, it was essential that the content of available resources can be expressed in Portuguese and in sign language. Photos, games drawings and field work were accompanied by an interpreter of the Brazilian sign language. The teaching sequences were developed to meet the needs of a class with deaf and hearing pupils with the theme, "The Urban Environmental Problems: From the world to my school".

## 22.2 Games, Pictures and Cartography

This work is not aimed to explore all the central elements of cartography (such as projections, scales, coordinates and mathematical relationships). Rather, this research considered cartography a tool to enhance the knowledge of the world and the living space of the pupils. For Callai (2005), a form of scanning the world is realized through the reading of space, which reflects the daily activities of men in their environment. Thus, reading the world goes far beyond reading cartographic representations which reflect the local area, sometimes distorted because of the cartographic projections adopted. To read the world is not just doing a reading of the map or by the map, although it is very important. You do the reading of the living world, built daily, expressing both our utopias and the limits that we are put in, either in the scope of nature or within the context of society (cultural, political, economic) (Callai 2005).

Therefore, in the development of activities with deaf pupils, the use of maps is not the sole factor, but is an important visual resource for exploring the living space and the teaching of geography. By corroborating this view, it is said that in the school, the discipline of geography, in particular the contents of Cartography, offers subsidies to expand the pupils' understanding of the place where they live and work (Juliasz et al. 2007). Cartography is, therefore, an important resource for the teaching of Geography, because it allows the representation of different perspectives of space and scale that is appropriate for teaching. Thus, the mapping allows the pupil to understand the manner in which he or she is inserted in space, which can be at local, regional and global levels. Through maps, they will be able to distinguish the different and distant locations, giving them the possibility of a more critical view of reality that they belong to.

According to Almeida (2001), the importance of learning in a socio-cultural context of modern society, as a necessary tool in people's lives, requires a certain mastery of concepts and references to spatial displacement and ambiance, and more than that, so that people have a conscious and critical view of their social space (Almeida 2001). As Almeida and Passini (1989) point out, the living space refers to the physical space experienced through movement and displacement. It is learned by children through games or other forms by going through it, enclosing it, or arranging it according to their interests. Combining the use of Cartography, drawings, photos and games, we aimed to improve the understanding of space by pupils. In the case of deaf children, it was essential to value their vision. The option of showing and building drawings with them is in agreement with the studies based on the perspective of Oliveira Jr, who says: *"The drawing was really a scape option. Fleeing from the word, whether oral or written, as the only transmitter of knowledge and information. But it was also an option for that approach. Approaching to a language more suitable for the transmission of knowledge about space, where the elements would be presented spatially, without the need for a chain of words and expressions. When looking at a drawing we already have a global view of it and we can 'read' it in several ways, from various points. It is also like this with the space and the city"* (Oliveira Jr 2006). Regarding games, we shared Albres' idea (2010) that tells us that childhood is a significant period for language acquisition. At this stage, recreational activities are built in the form of games. The same author points out the relationship among games, language and cognition development in deaf children. For this reason, during the study, recreational activities were privileged that would allow communication with the pupils, and that would give importance to the visual field.

## 22.3 Research

### *22.3.1 The Study Area*

The research was conducted at the Municipal School Professor Armando Grisi, located in Jardim Paulista, in the outskirts of Rio Claro, São Paulo, Brazil, housing a community of low income. This explains why the neighborhood has several problems in the urban environment. A study on the perception of space and character of the "environmental educator", which embraced and integrated the pupils in this school, proved to be of great importance. Even in the school where the research was conducted in 2006, we started work only with the inclusion of deaf pupils in the district where it is located. Because of an imposition policy, the school received these pupils. However, the support was poor. The school principal in the period in which the proposal was implemented (on a visit made in 2009) reported that, despite the difficulties, staff and pupils committed themselves to actually include the deaf community in the context of regular school. From this context, they took their own project and structured (bilingual), defined needs, strategies and possibilities. According to the director, this work would be impossible if there was no engagement of all staff and pupils. Currently, City Hall/Department of Education provides funds to guarantee an expert educator in BSL (Brazilian sign language) accompanying nine deaf pupils who attend school regularly in elementary school. Regarding the distribution of deaf pupils in classes, the Director during the visit in 2009 told us that at start, all who attended the same grade would be placed in the same class. However, she said, motivated by the interests of hearing pupils to learn sign language, the classes were rearranged so that there was a greater "integration" and learning from each other, even though this might be discussed.

### *22.3.2 Pupils Involved*

Initially, when this survey was idealized, we thought of the implementation of activities in classrooms with deaf and hearing pupils. However, during the visits and conversations with the teacher and pedagogical professional hired by the school, it was decided to carry out activities only with the deaf pupils. We believed that this way the main design goals would be achieved. It is worth noting that all activities were planned and designed so that they could be developed in classrooms with deaf and hearing pupils. At the same time, the different levels of proficiency in Portuguese and in the written form of Brazilian sign language, the different levels of deafness as well as the various difficulties of the pupils involved eventually constituted a very heterogeneous and mixed group. In some activities, we divided them into two groups with the intention of making a careful observation and rethinking the attitudes. The pupils involved in the project were nine children ranging from 9 to 12 years old. All these pupils were deaf in elementary school and in regular school.

## *22.3.3 Activities*

### 22.3.3.1 Building the City

The first practice with the pupils sought, from the appreciation of the visual field and the Brazilian sign language, awakening the pupils to the impacts of Brazilian urbanization and environmental problems resulting from this process. Starting with the practice of assembling a jigsaw puzzle containing a drawing of a modern city (Fig. 22.1), we tried to stimulate the visual discrimination, analysis, synthesis, and visual-motor coordination (Albres 2010). The puzzle was designed using cardboard paper, the drawing of a city, and impermeable adhesive paper to make it more resistant. If we had had more time, it would have also been possible that each pupil

**Fig. 22.1** Pupil building the city

would be asked to design their own puzzles with their own drawing and then be exchanged between them. After the pupils managed to form the puzzle, they were asked to locate the elements such as cars, buildings, factories, trash, etc.

The next step of this activity was a sequence of pictures, which sought to portray "from the natural environment to urban problems", presenting situations such as the construction of the first railroad tracks, the industrialization of the cities, the verticalization, the queues for jobs, socio-spatial segregation, visual pollution of cities and others. In all images shown, the subtitles were included in Portuguese and sign language. Ending this practice, it was proposed that the pupils would draw what drew more attention to them. Some pupils drew some garbage in cities, buildings, others drew up the slums. Interestingly, one pupil chose to draw a rural landscape, arguing that it would be difficult work for him, because there were many different urban elements. This showed that some situations can indicate misunderstanding, which may actually be revealing. With this practice, it was clear how the language is fundamental in the organization of thoughts: those who had a more structured language had an easier way to understand the subject matter and then represent it.

Regarding the designs produced by pupils, it was noted that at first (Fig. 22.2) the pupil was concerned with the aspects of designing an urban landscape, a building and a vehicle, undeniable "symbols" of cities. In this second design (Fig. 22.3), it is very interesting to note that the pupil took care to show her mastery of Portuguese in the written form when writing garbage in the trash, where a kid throws a package. We also noted the contrasted ideas of two different situations: a guy throwing the

**Fig. 22.2** The design of the building and truck

**Fig. 22.3** The representation of opposing situations

trash from a car and another having a more correct environmental point of view. Still related to the design, we also note that the pupil can recognize and draw the elements into perspective, a characteristic usually developed after 8–9 years of age (Almeida and Passini 1989). Pupil Nathalia soon was encouraged by her family, which is the listener, to learn the Brazilian sign language. She has no difficulties to realize the contents exposed, moreover, she is one of the more attentive pupils. In the next representation (Fig. 22.4), the pupil exhibited some situations also presented and discussed by the images shown previously. The pupil portrays a girl with some bags in her hand, perhaps inspired by the images on consumerism that were also presented to them. The rain and the house might want to tell us something related to floods. However, it is worth noting that the pupil does not place the drawing elements at the bottom of the paper. For a child of 11 years, could we say that he/she is pretty childish for his/her age? Can we infer about their cognitive and spatial sense?

In the last drawing (Fig. 22.5), there is no doubt that it makes us reflect on different aspects. The pupil has serious problems regarding the acquisition of the Portuguese language in the written form and according to the interpreter, presents poor and has difficulties with BSL. When he finished his drawing, we questioned him about what he had done. He told me that he had designed a "hill" (note the square at the bottom left, which would be boxes) and a car "falling" into a flood. It would be fair to say that their language deficits interfere with their ability to represent it? Did the absence of a structured language trouble him to express what he really liked to do?

Fig. 22.4 "Consumerism rain and the house"

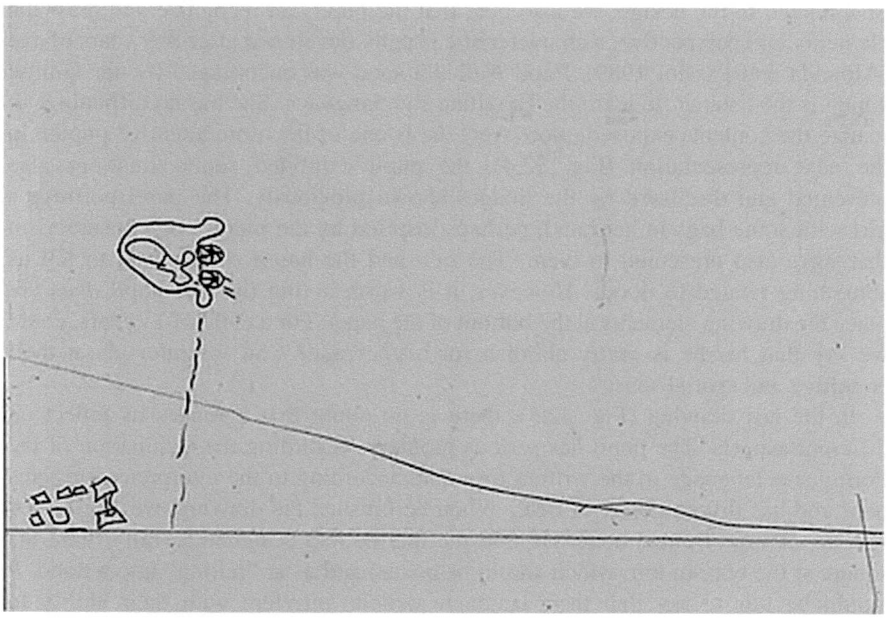

Fig. 22.5 The suburbs and the car

#### 22.3.3.2 From the World to My School

The second stage of the project came up with the proposal of giving the pupils an idea of Geographic Space before talking about environmental issues. We used a globe and different maps, including one of our cities, Rio Claro, as our idea was to give the pupils a global and a local idea of space (Fig. 22.6). Having found where Brazil is located in the globe, they were given a colorful map of Brazil printed in A4 size in order to check if they were able to identify which country it was. This activity aimed to encourage them to acquire Portuguese vocabulary in the written form, and present them important elements of maps and charts (the separation of title, content and caption). The map of the political division of Brazil was given after that, and they were supposed to paint only the province in which they lived. The map was printed without color, leaving all the states represented with white background, so that pupils could paint our state. First we provided them with the globe and the physical map, and then gave them the political map that aimed to show them that space is not "born" with boundaries. The idea was first to introduce

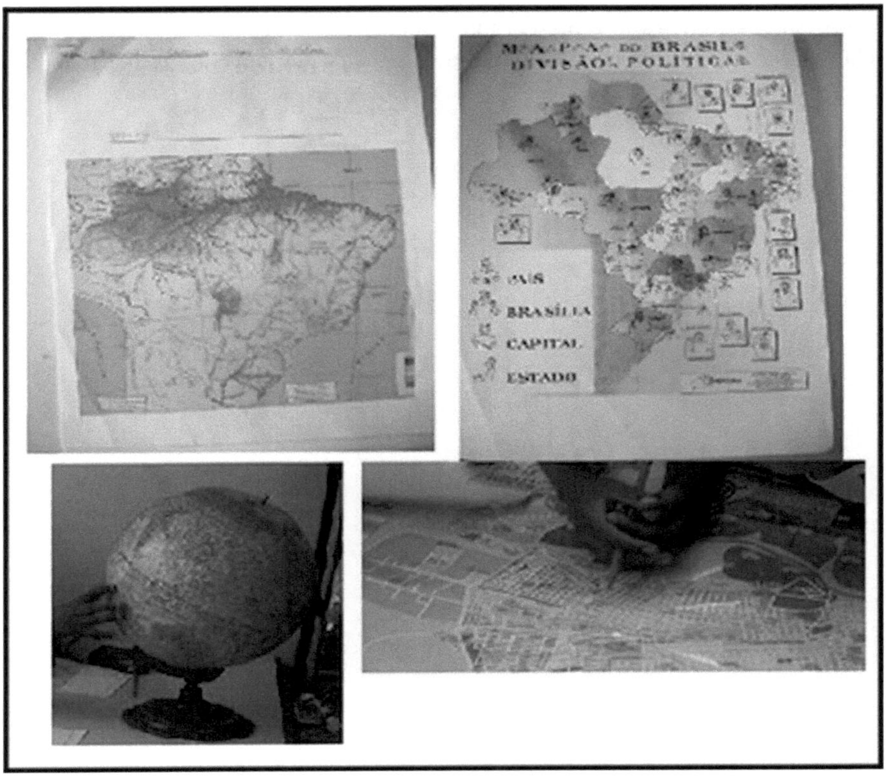


**Fig. 22.6** Materials used in this activity

the pupils to a unique Brazil, explaining them the differentiation of colors depending on the types of relief etc. We wanted to show them that the States are a recent chapter in Humanity's history, as pointed out in the "Notes on the Education of visual maps: The nature of the idea of representation", by Oliveira Jr (2009). The next step was to get the map of the city of Rio Claro to try find out the localization of the school and of the houses in which each of the pupils lived. This phase of activity was a lot of fun to the pupils because they all wanted to be faster than the others in finding the school and their homes. For this activity, we used only a city map available at newsstands. To finish this practice, we individually asked the pupils to draw up mental maps of the possible routes from their homes to the school. Some of them went further by placing signs in Portuguese language about reference points along the way, like stores, streets, etc. By doing this activity, we aimed to increase their spatial perception in different scales and we wanted to encourage the pupils to think about a possible hierarchy: World—Continent—Country—State—City—Suburb.

On this first map (Fig. 22.7), made by the same pupil that made Fig. 22.2, there was a concern to make it flat, like a conventional map. It is interesting to note that again she was concerned about indicating reference points in Portuguese language (aunt's house, bar, market, etc.). The pupil also used symbols to point her reference marks, such as crosses to represent the churches along her journey from home to school. One aspect that caught attention was that while developing her map, the

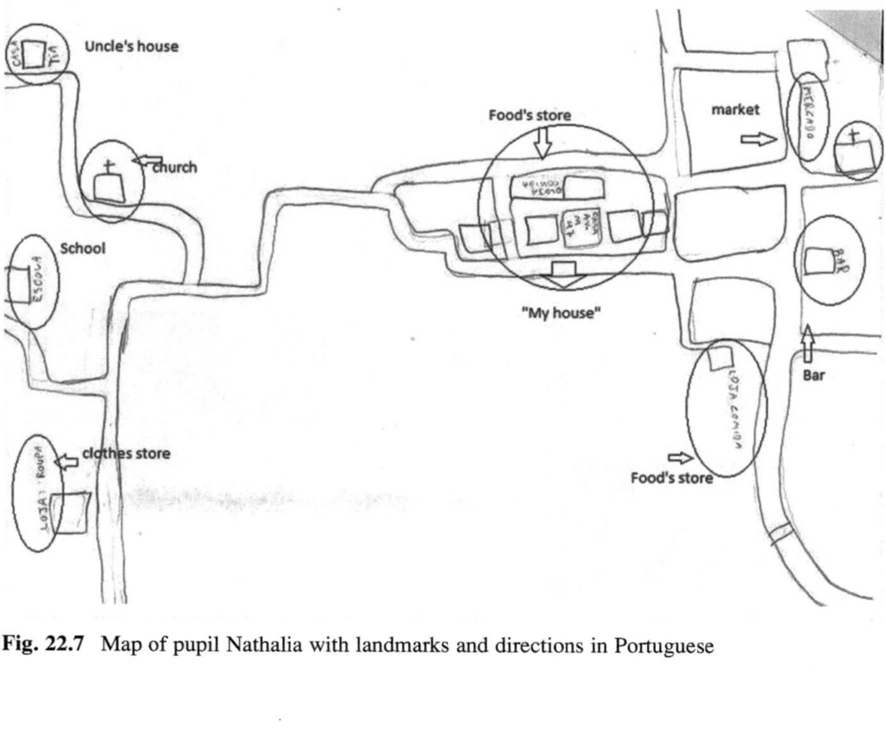


**Fig. 22.7** Map of pupil Nathalia with landmarks and directions in Portuguese

pupil told me that she had to draw everything "too small" in order to fit everything in there, which gave me the impression that her sense of scale had already improved.

In the second map (Fig. 22.8) drawn by another pupil, only a reference point, that is another school was indicated, which is located in the same street where the school they attend is. In this case, they are two children of the same age, but the presenting levels of perception and spatial representation are quite different. We can even say that while the previous representation has the characteristics of a map—locating the places in two-dimensional plane, trying to place their way and streets (Marandola and Oliveira 2007)—the other one (Fig. 22.9) can be considered a design. In it, we noticed the lack of perspective by portraying the street seen from above and the other elements seen from the front (Almeida and Passini 1989).

De Paula (2010) warns that the problems regarding the mental maps are concentrated, in part, at the time subsequent to the acquisition of material but before data analysis. She says that both representations have their advantages and disadvantages. A two-dimensional mental map is advisory because the individual fits into a system of directions (right, left, front and back), but loses ground to the possibility of developing related images, such as topography. The oblique representation from a skyline, despite decreasing the clarity of spatial relationships between Euclidean distances and directions, allows the individual to draw the elements according to their size and shape. It is this perspective that we often see the world. There is no restriction on the form of representation (De Paula 2010).

We also highlight that in his drawing, Gustavo was careful to indicate the room number where we gathered to do the activities. He possibly showed his own body as a reference for the location of objects, possibly not having gone through the process of "decentering", which according to Almeida and Passini (1989) is the passage of child egocentrism to a more objective approach to reality through the construction of conservation structures that allow children to have a thought more reversible. This is because they begin to consider other elements for the spatial location and not just their perception or intuition about the phenomena.

The third map was made by the same pupil who made drawing 4. It was clear that he had difficulty in representing his perception. We are not saying what is 'right or wrong', especially because the maps and drawings are mainly expressions of their creativity, but we cannot deny the difficulty shown by this specific pupil. In general, we can say that this activity with drawings and maps was really informative about the characteristics and needs of the pupils.

#### 22.3.3.3 Fieldwork in the School District

In order to aggregate the elements discussed in the previous practices, we decided to do a fieldwork on the block and around the school. The area around the school has serious problems of urban and environmental policy, such as unpaved and poorly

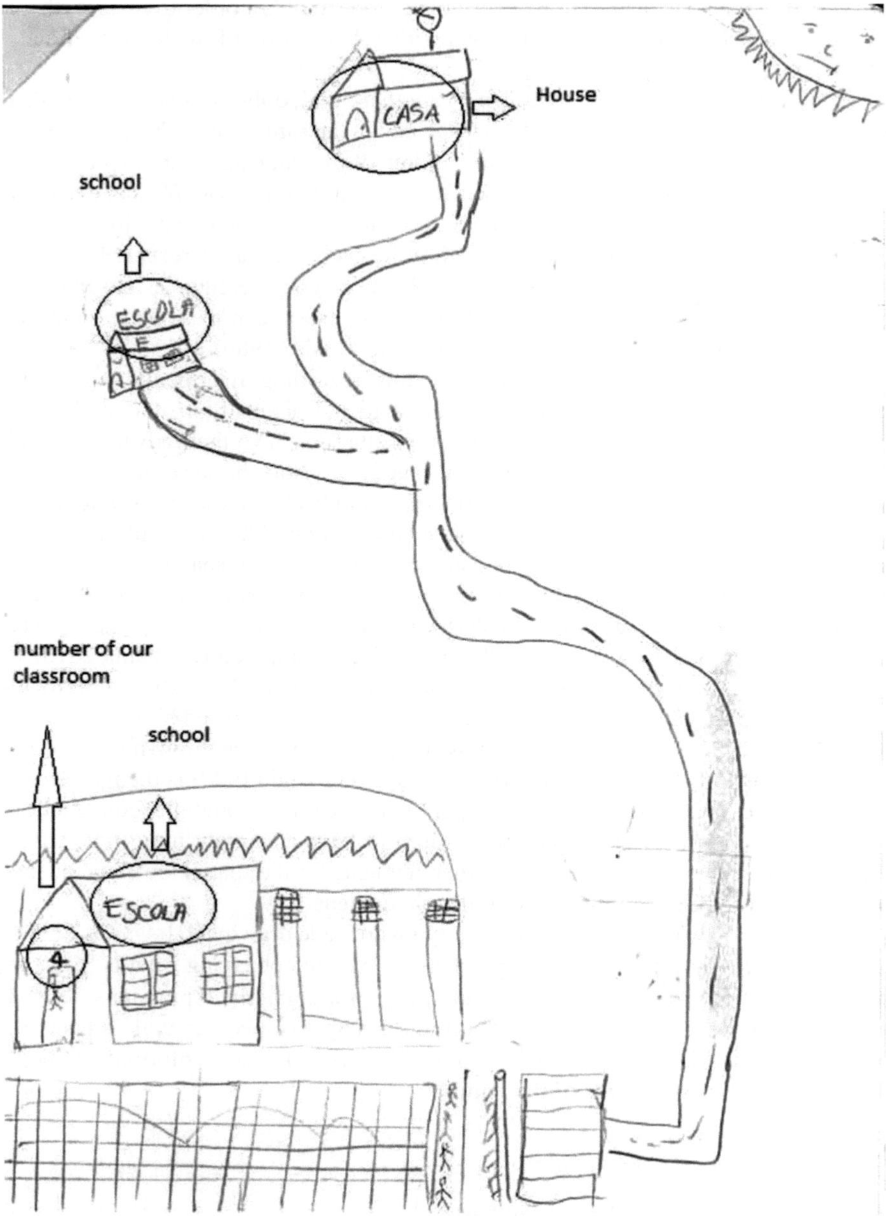


**Fig. 22.8** Representation made by pupil Gustavo

maintained roads, garbage dump on the banks of a small stream, fires and illegal housing. In order to develop a complete task, before starting it, we returned to the idea of different scales of spatial analysis via the image below (Fig. 22.10).

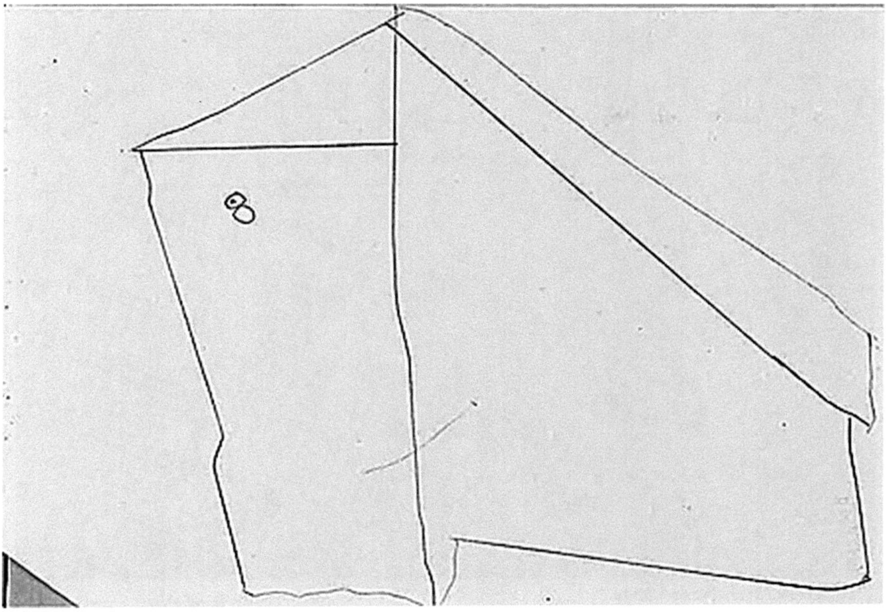

Fig. 22.9 Representation made by pupil Paulo

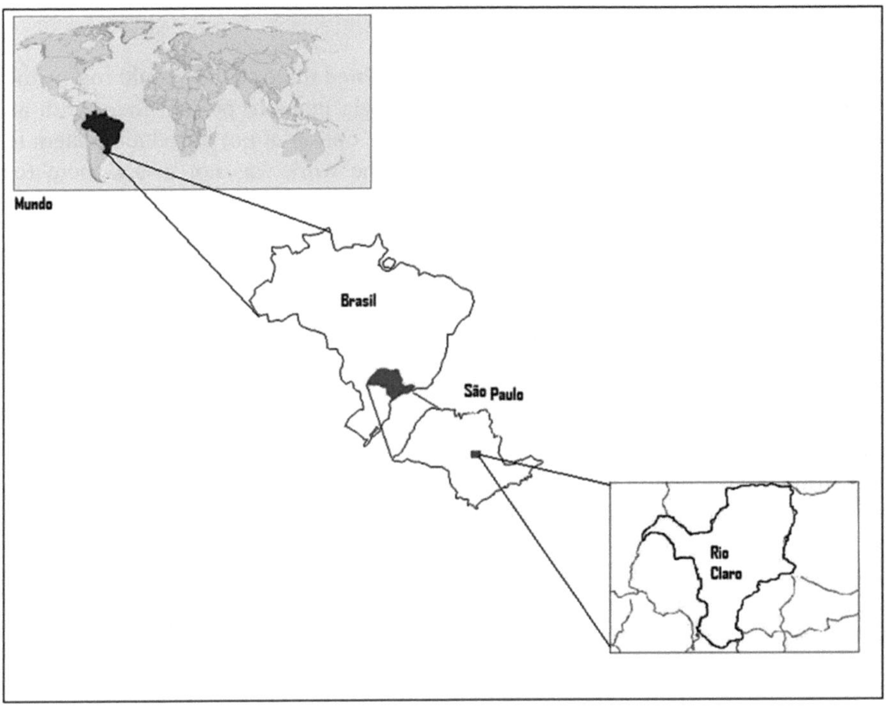


Fig. 22.10 From global to local (Prepared by Tiago Salge Araújo, 2011)

**Fig. 22.11** Map delivered to pupils with the course of fieldwork

We provided maps to the pupils with a highlighted route so they could follow the steps of the fieldwork. We decided to use Google maps to present them with an accessible and free use of maps (Fig. 22.11). The option of not introducing them to 'scales' was made because the main goal of the work was not to ask them for 'calculations' and cartographic operations. Along the way, besides the map, each pupil had a digital camera, so they could record the aspects that drew most attention (Fig. 22.12). Besides enhancing the pupil's visual field perception, the idea was to capture images able to allow a further analysis of the elements that were more interesting to them. This activity was more important for the deaf pupils. Helped by a school interpreter, we talked about the importance of the preservation of our water resources as well as about the impacts of the urbanization process and of the irresponsible garbage in the Corumbataí river, which is located near the school that they attend.

During the fieldwork, besides the problems related to the waste and pollution of water sources, housing issues also attracted the attention of pupils. Back to the school, we had a picnic in the schoolyard and talked about the impressions the pupils had about what they had seen. They liked the beauty of the sky, the animals they found along the way, such as cows, birds and dogs; they also liked the fieldwork in itself, but they did not like the garbage, the dirty water and the holes on the road.

Fig. 22.12 Some pictures taken by the pupils during the fieldwork

#### 22.3.3.4 From the Reality to the Representation

As a final activity, it was proposed to build a model of the route taken during the fieldwork with the pupils (Fig. 22.13). This activity focused on the importance of stimulating the pupils' perceptions of the living space. According to Freitas et al. (2008), the use of models allows the representation of landscape elements in three dimensions, providing a synthetic model of the complex reality of the use and occupation of urban land. Quoted by the same authors, Filett (2004, as cited in Freitas et al. 2008) points out that the reduced models bring children to the materialization of real spaces that provide concepts often not understood by them, since the children of the first cycle of elementary school have a level of abstraction in development and require viewing to understand them most of the times. Certainly, the collective construction, along with pupils, is the most appropriate way to explore all aspects and steps of a model.

According to the authors mentioned below, the model can be used as a collective representation of space, which explores the dimensionality, therefore helping to improve the spatial reading of each individual through the exercise of viewing and interpretating the city (Freitas et al. 2008). However, in order to conclude the research

**Fig. 22.13** Model of the route taken in the fieldwork

on time , the construction of the model was made by me and Carla Lombardi, the school interpreter. The pupils then only used and explored it. In order to give them a sense of the land level, the model was based on a 1:25,000 topographic scale of the area. One half of the model would be the representation of the trajectory and the other half would be a legend in 'Brazilian signs language' and in Portuguese. In order to stimulate the implementation of the "real to the abstract", some photos taken by the pupils themselves were printed and glued on small polystyrene and sticks, so they could post them on the model. Each photo was accompanied by a title in Portuguese and in "Brazilian signs language". Moreover, numbers were given to the pupils and they had to put them in the model in the most appropriate place. This aimed to encourage the pupils to look into their memories for the location of the elements observed in the fieldwork (real) and transport them to the model (abstract).

In this activity, all pupils participated in the survey were involved together and the activity lasted for approximately 2 h. When we discussed the application and handling of the model, we thought that it might be best to recall them from the pictures they took, the route taken during the fieldwork. However, to our surprise and delight, it was not necessary. As soon as they saw the demo, the pupils were very curious and wanted to know everything, "whether it was the school, whether it was the neighborhood, etc..." We then began to work with each element; we asked them to look at the legend and to try to identify the elements. Then we asked them about the relationships by putting the number indicated in the legend (Fig. 22.14) in

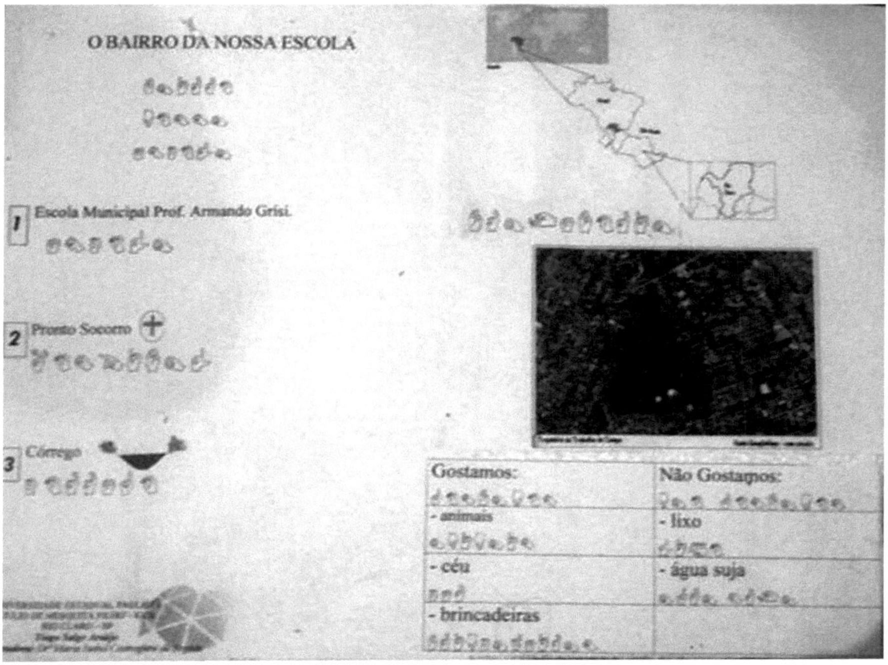


Fig. 22.14 Large bilingual legend containing references and a list of points that the pupils liked and disliked in fieldwork

place in the representation. Pupils Renato, Dalila and Nathalia found the most suitable places soon, while the others were sometimes distracted by small details of the model, such as the composition of materials.

Next, we asked them to fit the photos (with subtitles BSL) with the most appropriate place (Fig. 22.15). The photos chosen were those shown in Fig. 22.14. Again, some pupils excelled showing a greater interest, while others just participated. Finalizing the framework, we discussed this activity: "like—don't like". In a general evaluation of this activity with the model, we conclude that it would have been the ideal way if the activities had been done individually or in pairs. Thus, the number of negative answers would probably have been smaller and everyone would have been involved in a similar way. However, we understand this as an excellent activity as well as a method of introducing the elements of cartography, allowing visual discrimination, memory and perception of living space.

## 22.4 Final Considerations, Far from One Conclusion. . .

Start by saying that methodological and practical issues are far from a model. This is because if the deaf community shares cultural and linguistic elements, which enables us to define them as such, on the other hand the specific needs and

**Fig. 22.15** Photos posted on the model

characteristics of each subject and pupils, it is impossible to draw a homogenizing circle. If the listener in a culture has the same logic, do we recognize the differences among pupils? I mean that we should not expect formulas already set for teaching and we can work with deaf pupils and/or mixed classes. Each pupil, deaf or listener, has his/her potential, specific needs and desires. With respect to all the research and activities, we were very clear how the language is the key element when it comes to teaching the deaf. The respect for their language and their appreciation of the visual field may be the only thing we can say categorically. It is impossible to think about full inclusion if the classes are designed under the logic "ethnocentric listener" (Fernandes 1998).

The teacher, who is faced with this situation, should dip in a constant process of reflection and their teaching practice. If we want a democratic school, we should expect it to be prepared to receive each of the pupils, deaf and hearing. The teacher of deaf pupils should contact the relevant bodies and dedicate him/herself to know this other culture. The teacher prepared for this situation must live with the deaf community, and seek, where possible, to know more and learn sign language (even if the school relies on an interpreter) and think about broader pedagogy (Skliar 1998 as cited in Strobel 2006).

In the context of cartography, this proved to be an important instrument of perception and the knowledge of living space. Through and with it, we awaken in pupils a greater sense of space, reflecting on our position in time and space. The first activity with the puzzle, photos and drawings, allowed us to learn more of the pupils and understand their forms of representation. From the second activity with maps, globe and maps, we believe that we were able to show them important elements of cartography, also indicating their utilities and various forms. The fieldwork and handling of the model enabled a performance analysis and spatial perception rather interesting, transposing the real to the abstract, directing and "freezing stares". In general, we see an evolution in the spatial perception of pupils, comparing the first activities with the activity of the model. We therefore believe that the objectives initially proposed by the project were achieved: we can prepare materials in cartography, in Brazilian sign language and Portuguese, which enabled the pupils got be involved, to expand their knowledge of the mapping applied to environmental studies, to begin studies using satellite images, aerial photos, letters and fieldwork. Moreover, the activities provided an insight to the lived space and environmental issues involving water resources in the study area.

We are sure that the work is not completed, now depending on other professional teachers to continue encouraging and developing skills with these pupils. Perhaps the key to the question is whether we realize that "the difference is us" (Magalhaes and Stoer 2005). According to these research proposals, we should look at the difference in another perspective: This discharge of voice may not escape the boundaries of another as an object, this time through our political generosity. This is why we put ourselves under the statement "the difference is ourselves". The difference in this perspective is the "*product of a relational game in which there is no longer a privileged center from which one can determine who are the others, who are different*" (Magalhaes and Stoer 2005). We share the ideas of this authors, when they say that the other is different and so are we. The difference is the relationship between different. We therefore conclude by saying that when it comes to "special" or inclusive education, we must look at all those involved as "different" and not just the "other". The search for dialogue and promotion of teaching that really reaches the "other" should be the ultimate goal. With respect to Geography and Cartography, we proved that there are ways to do this; it behooves us to go on researching and looking for ways to promote a truly inclusive education.

# References


Albres NA (2010) Surdos & Inclusão Educacional. Arara Azul, Rio de Janeiro

Almeida RD, Passini EYO (1989) Espaço geográfico: ensino e representação. Contexto, São Paulo

Almeida, RD (2001) Do desenho ao mapa: iniciação cartográfica na escola. São Paulo: contexto, 115p

Bogdan RC, Biklen SK (1994) Investigação qualitativa em educação. Porto Editora, Porto

Callai HC (2005) Aprendendo a ler o mundo: A Geografia nos anos iniciais do Ensino Fundamental, vol 25. n. 66. Cad. Cedes, Campinas, pp 227–247

de Paula LT (2010) Mapa mental e experiência: um olhar sobre as possibilidades. XVI Encontro Nacional de Geógrafos, Porto Alegre

Fernandes SF (1998) Surdez e linguagem: é possível o diálogo entre as diferenças? Dissertation. Universidade Federal de Santa Catarina, Florianópolis

Freitas MIC, Lombardo MA, Ventorini SE (2008) Do mundo ao modelo em escala reduzida: a maquete ambiental como ferramenta de formação do cidadão. In: Mercator, vol. 12. pp 90–102

Juliasz PCS, Freitas MI , Ventorini SE (2007) Proposta diferenciada de elaboração de mapas táteis. In: XXIII Congresso Brasileiro de Cartografia e I Congresso Brasileiro de Geoprocessamento. CBC 2007. Sociedade Brasileira de Cartografia, Rio de Janeiro, pp 2475–2483

Marandola JAMS, Oliveira L (2007) Desenhos e Mapas: representações e imagens do Urbano. In: XI Encontro de Geógrafos da America Latina—XI EGAL, Universidade Nacional de Colômbia, Bogotá

Oliveira Jr WM (2005) A produção da escuta a partir de imagens. In: Anais do 8 Encontro Nacional de Prática de Ensino de Geografia, 1–25. Dourados, MS

Oliveira Jr WM (2006) Desenhos e escutas. In: 29ª Reunião da Anped, GT 12. Caxambu

Oliveira Jr WM (2009) Apontamentos sobre a educação visual dos mapas: a (des-) natureza da ideia de representação. In: Colóquio de Cartografia para Escolares. UFJF, Juiz de Fora

Skliar C (Org.)(2001) Educação & exclusão: abordagens sócio-antropológicas em educação especial. 3. ed. Porto Alegre: Mediação

Stoe S, Magalhães, A (2005) A Diferença somos nós. A gestão da mudança social e as políticas educativas e sociais, Santa Maria da Feira Edições Afrontamento

Strobel KL (2006) A visão histórica da in(ex)clusão dos surdos nas escolas. In: Educação Temática Digital, vol 7. n. 2. Campinas, pp 244–252

Tiago SA (2011). A Cartografia nos estudos do meio ambiente em classes mistas: por uma prática bilingue visando a inclusão dos alunos surdos. Trabalho de conclusão do curso de Geografia. Universidade Estadual Paulista,UNESP, Rio Claro/Brasil

# Chapter 23
# Study on the Acquisition of the Concept of Spatial Representation by Visually Impaired People

**Silvia Elena Ventorini and Maria Isabel Castreghini de Freitas**

**Abstract** This article aims to present the results and analyses of research whose main objective was to investigate how visually impaired people are able to learn and draw as well as what the importance is of acquiring the concept of spatial representation for reading, interpretation and analysis of tactile cartographic documents by this segment of the population. The data and analyses presented in this paper were collected from a Special School and a Children's Rehabilitation Center. The research concluded that a blind child develops this concept in the same way as any other child: they acquire the concept of the permanent object, acquire semantic memory and graphic act, attribute sensory and physical meaning to the graphic act and show difficulties to draw locations and objects that do not possess significance in their experience of life.

## 23.1 Introduction

Probably every teacher of the lower grades at Elementary School has already observed a pupil in the act of drawing and seen graphic shapes materialize on the paper similar to those of other children. Through imitation and mediation of adults or other children, children without visual impairments are encouraged to develop graphical representations from the earliest age. In their act of drawing, they produce similar graphic forms on the paper, usually considered "stereotypical": the sun represented by a circle with rays emerging or set the house represented by a triangle on top of a square.

S.E. Ventorini (✉)
Universidade Federal de São João del Rei, UFSJ, São João del Rei, MG, Brazil
e-mail: sventorini@ufsj.edu.br

M.I.C. de Freitas
Universidade Estadual Paulista—UNESP, Rio Claro, SP, Brazil
e-mail: ifreitas@unesp.br

L. Zentai and J.R. Nunez (eds.), *Maps for the Future*,
Lecture Notes in Geoinformation and Cartography 5,
DOI 10.1007/978-3-642-19522-8_23, 

These drawings have been the subject of studies by many researchers, both nationally and internationally. Relevant research on how drawings express the children's perceptions of the living space as well as understanding of the mental and structural factors responsible for the process of producing these graphic signs contribute to the development of methodological procedures that assist in expanding the cartographic and geographic concepts of these individuals.

However, note that there is insufficient research on the drawings by blind subjects and how they can express knowledge about a particular object or environment by means of graphic language. Many parents and teachers believe that the act of drawing is only possible through the use of a visual channel. For this reason, they try not to encourage blind children to draw. Others (teachers) who teach blind children to draw expect them to produce representations similar to those of children who can see.

In this research on the spatial organization of a group of blind pupils, we engaged them in an activity where they were to render drawings. This option was considered when we observed that the pupils produced graphical representations of isolated objects. As a basis, we also used the reflections of the authors of School Cartography about the importance of drawings for children to learn about standard maps and about how these children's drawings are representations of their thoughts about the world (Almeida 2010). We also sought to investigate if drawings have the same importance for the blind. The study referred to on spatial organization was published in a book by Ventorini (2009) entitled, A experiência como fator determinante na representação espacial da pessoa com deficiência visual.

However, there were unanswered questions in the aforementioned paper: how, for example, these pupils, particularly one who lost his sight in the early years of his life, acquired the necessary concepts to express themselves through drawings? What are these concepts and what is the process to acquire them? Why are not all blind people able to draw, as we observed in theory and practice involving a blind pupil, who was 15 years of age?

In academic publications we discovered that drawing activities do not make up part of the everyday lives of blind people, perhaps due to the fact that it is a visual activity. In our view there is no difference in the meaning of drawings produced by blind children and those who are not if they are interpreted in the following way:

> A child's drawing is, thus, a system of representation. Not a copy of objects, but an interpretation of that which is real, done by the child in graphic language. Looking at a drawing in this way, it is possible to see beyond the infantile stages of the drawing, and analyze them as an expression of a language, which children appropriate to make their impressions visible, thereby socializing their experiences. In one representation, "X" does not equal reality "R", which represents this, and the connection can be either analog or arbitrary (author's emphasis). The drawing establishes an analog link with the object represented, as the visual signifiers are of the same nature as its meaning (our emphasis). [...]. The graphic image is not, therefore, a copy of reality. It depends on the systems of representation and its perception of the object and its graphical abilities. (Almeida 2010)

The considerations of the author, although they concern the drawings of children with normal vision, also apply to the drawings prepared by the group of blind pupils, who participated in our survey. Their drawings are representation systems, not copies, they function as a means of socialization, as they are visible to the eyes of those who see them and visible to the touch of those who feel them.

Considering that for a blind child to render graphical representations, it is necessary to acquire the concept of symbolism—that something can represent another object or the same object –, how does such a child develop this concept and represent it through drawings? The question is relevant, because it questions the association between seeing and knowing and refers to the questions made by Batista (2005): "*What is knowing? To see is to know? Sensory feeling is to know? One of the responses current in psychology and in the educational environment relates to the act of knowing to the acquisition of concepts.*" This answer takes us back to the question: how does a blind child develop the concept of symbolism and represent it in drawings? Another question that crossed our minds was: could it be that when a blind person learns to draw, that this facilitates the acquisition of concept mapping?

Thus, the purpose of this paper is to present findings and analyses of the research developed, whose main objective was to investigate how blind people can learn to draw and how important the acquisition of drawing concepts is for reading, interpretation and analysis of tactile cartographic documents among the members of this part of the public.

The data and analyses presented in this publication were collected from a Special School, whose name is EMIEE Maria Aparecida Muniz Michelin – José Benedito Carneiro : Deficiente Auditivo e Deficiente Visual located in Araras, a city in the state of São Paulo, from five pupils who participated, and from the Centro de Reabilitação Infantil Princesa Vitória located in Rio Claro, state of São Paulo, in 2009, involving a 15-year old ninth grade pupil, who has been blind since birth.

## 23.2 The Context Where Blind Pupils Learn to Draw

At the Special School, the material used for drawing consisted of a board covered with thin fabric, crayons and A4 150 gsm paper. The fabric was attached to the back of the board by small tacks. One of the edges of the paper was attached by a clip, which was part of the board. When scribbling on the paper with the crayons, the thin fabric produced a high relief that could be felt by touch.

After drawing the first lines, these blind pupils found a lot of pleasure in using this material and being able to feel by touch what they were scribbling. There were other variables that allowed these pupils to continue with their drawings, as discussed throughout this document, but the satisfaction of the pupils upon feeling the effect of the movement of the crayons on the paper was evident.

The same was not observed at the Centro de Reabilitação Infantil during the collection of the data. The drawing activities were not included in the blind pupil's daily routine. The facility also lacked suitable material for this purpose. For the drawing activities a sheet of aluminum and a pen were used. After scratching a mark in the aluminum, the line produced a low relief on the side where the drawing was being rendered and a high relief on the opposite side. So to feel by touch the lines etched into the sheet, the pupil had to run their fingers over the opposite side of the drawing (Figs. 23.1, 23.2).

In the first activity with this material it was possible to observe the pupil's discomfort in using a material that was so different from that of his colleagues, as well as a certain lack of interest in the material, which did not allow him to effectively feel the movement of the pen. This material was replaced with a board covered with thin fabric, the same that was used at EMIEE.

Figures 23.1 and 23.2 show that the pupil does not have the motor coordination necessary to etch a scratch on the material, nor the coordination to "see" the lines drawn, even when he felt the opposite side of the sheet on which the lines were in relief. We also point out that even when providing suitable material for this pupil to draw on, the results collected indicated a lack of manual motor coordination. The lines produced by the pupil did not stand out enough to be felt by touch and showed up very faintly on the paper. We put it down to a lack of bimanual coordination and, consequently, of graphic memory.

**Fig. 23.1** Material used for drawings at the rehabilitation center

**Fig. 23.2** Pupil feeling the lines on the reverse side of the sheet (Source: collection of the tactile cartography group–UNESP; photo: Juliasz (2009))

## 23.3 Bimanual Coordination and Graphics Memory

In the book *Discapacidad visual y destrezas manipulativas* (Rubayo et al. 2007) are suggested activities and materials that aid in tactile stimulation. Many of the authors' suggestions were observed at the Special School, such as how to make balls of paper, paint with your fingers using open molds, classify and sort a mixture of small objects, differentiate textures, recognize miniatures and figures in high relief, recognize geometric figures, differentiate sizes of objects, etc.

The authors' suggestions are to increase the sensitivity of the fingertips, the pressure when closing and opening one's hands and fingers and bimanual coordination. The hand actions undergo a proper sequence of motions, and for Rubayo et al. (2007) the principles of bimanual coordination are:

- Handle delicate and fragile objects differently from durable ones;
- Consider that the neurological maturity in itself is not sufficient for the development of manual skills, because opportunities are necessary for the hands to move in various situations;
- Know that each hand has an asymmetric function, and one is dominant and is more commonly used to the handle objects and in other activities and the other acts as a helper;

- Consider that the manual coordination serves to improve the efficiency and dexterity of the hands. The hands should be relaxed for any manual activity. The tension in the hands restricts the neuromotor reflexes that must be produced, causing stiff and distorted movements, affecting the activity to be carried out.

At the Special School we noted that the activities for mastering bimanual coordination were conducted with the pupils who lost their vision early in life and those who lost their vision later. In our observations we found that pupils who developed the impairment in adulthood demonstrated more difficulties in coordinating both hands in the activities, such as the activity where they used their non-dominant hand to feel the outline of the drawing. Moreover, we observed the tension in their hands as cited by Rubayo et al. (2007). This was caused by the trauma of having lost their vision.

Many of the activities developed at the Special School were aimed at Braille literacy or learning. Simon et al. (1991) and Ochaíta and Espinosa (2004) point out that—in general—the visually impaired read with the index finger of the dominant hand and use the index finger of the other hand to guide them when changing lines. In children's literacy development, the authors observed that a child only uses one finger, returning along the same line to start reading the next. They also emphasize the need for the development of tactile sensitivity and bimanual coordination. The drawing activity requires similar coordination, as the pupil draws with one hand and uses the other for orientation.

In order to hold and draw the hand and fingers need to undergo a series of exercises. Duarte (2008) points out that Marc Jeannerod calls this preparation "pre-training of the hand", which assists in the correlation between opening and molding of the hand to pick up the object and feel the size and shape of the object.

Little by little the child's mental memory develops of how to hold a pencil and produce the primary symbolic representation, which along with speech will allow it to express its repertoire from memory. In children who see, this occurs by imitation and mediation. In terms of the blind child this will need to be stimulated, since the lack of vision prevents it from observing and imitating the act of writing and/or drawing.

In work involving drawings we observed the importance of this memory. A blind 10-year-old pupil, whose name is Laura, was going through a phase of tactile development when we met her, and as her field of vision before becoming blind was not sufficient to visualize the shapes of objects and draw them, her motor memory was not developed enough to render, for example, a square.

In our study, we did not apply methods to teach the pupil how to draw, but educational situations in which the pupil understands how three-dimensional representations (models) may represent locations in their daily lives. In these situations, Laura would feel these models that represented known environments, exploring and comparing the real objects with the representations, then produced drawings of these environments. In other situations she would develop mental maps to solve problem situations. These procedures were adopted for all those who participated in the survey.

Figure 23.3 was the first drawing of the classroom produced by Laura after feeling a model representing the place (Fig. 23.4). The drawing was done a year and a half after the pupil began attending classes at the Special School. In the analysis of the drawing it is evident, even though the pupil is not proficient in graphic shapes, that almost all the existing objects in the room were represented – except for three chairs and the teacher's desk. The analysis also indicates that Laura had difficulties to represent objects in their correct locations, indicating that she still has not mastered the spatial organization of the classroom.

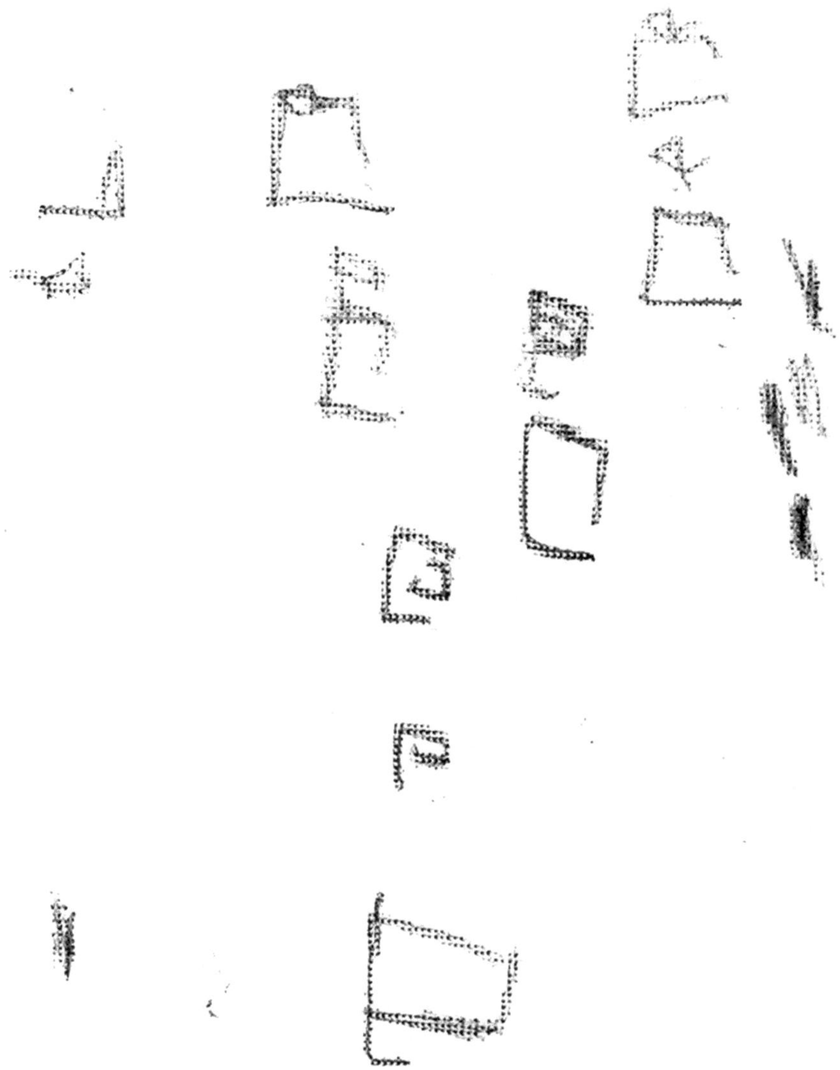

**Fig. 23.3** First mental map drawn by Laura

**Fig. 23.4** Model of the classroom

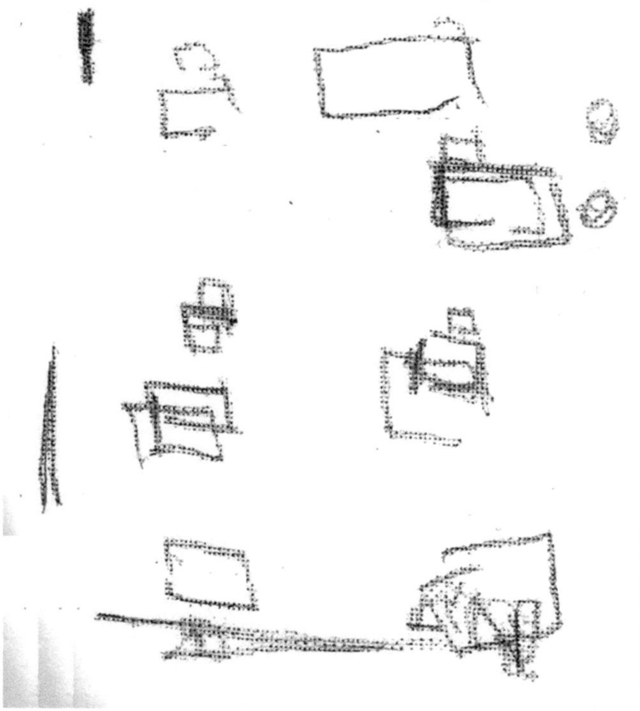

**Fig. 23.5** Drawing rendered by Laura after having explored her environment

Figure 23.5 shows the pupil's drawing after exploring the classroom with our mediation. In this mediation we explained to the pupil the position of the objects in relation to each other as well as their different sizes. When exploring the environment the pupil pointed out that the doll's house was not represented in the model, which she played with and that the pupil desks were not positioned one after another, as represented in the model. The exploration of the room was held 2 weeks after the previously reported activity and there was a change in the positions of the furniture in the room and we thought it was better not to change them to check if the pupil would notice the differences.

The pupil's drawing was done soon after exploring the locality. In this drawing we discovered that there was an improvement in graphic shapes, but confusion in the spatial distribution of the objects, especially with respect to the distance between the chairs and desks. The pupil represents these objects one on top of the other. In her representation of the dollhouse we are able to see that Laura sought to represent the slope of the roof by drawing a triangle, and a chimney with a rectangle. The shapes attributed to the details of the house were not obtained from tactile pictures, but through interaction with the object. The graphic knowledge used to represent these characteristics was acquired in the drawing activities of geometric shapes.

In extension activities, Laura built a model of the classroom where she attended classes at the Special School and then created a mental map of the site. There is evidence of the pupil's knowledge developing in this new drawing both in terms of the graphic and spatial organization. Figure 23.6 shows the model developed by the pupil and Fig. 23.7 shows the drawing.

Advances in spatial distribution of the objects represented were generated by the activities involving the model and by exploring the everyday environment. The improvement, however, in graphical shape is the result of the work done at school, that is, of the drawing teaching and learning situations, which have also improved Laura's motor memory for graphical shapes.

At the Special School the pupil was encouraged to represent anything he/she recognized, for example the shape of a circle, by a drawing. This drawing task was slow and the main objective was not to teach the pupils to draw the same as children that can see, rather it was to contribute to them having their own personal brand, as well as allowing them to understand geometric figures.

The movement associated with drawing on paper assisted in motor coordination and helped the pupil to acquire the necessary grip to hold a pencil and/or a pen and write his/her name. Laura's improvement in presenting graphic shapes resulted from the activities directed at developing her literacy. The shape used to represent the slope of the roof of the house is similar to the slope of the letter *A*, learned by the pupil.

When we look at how the pupil writes her name, beyond the concepts already discussed, it is evident that she needs to acquire the concept for rendering sloping lines to write the letter *A*, whose sloping lines are simpler when compared to the letter *N*. Moreover, the relationship of space between the letter and the relationship of size also require acquisition of concepts. When we looked at Laura's writing we found that there are not any significant discrepancies between the letters showing that the pupil acquired the necessary concepts to write his name, as well as the necessary motor coordination.

The same did not occur with another pupil, Bruno from the Children's Rehabilitation Center. When he was asked to draw a mental map of the computer room, where

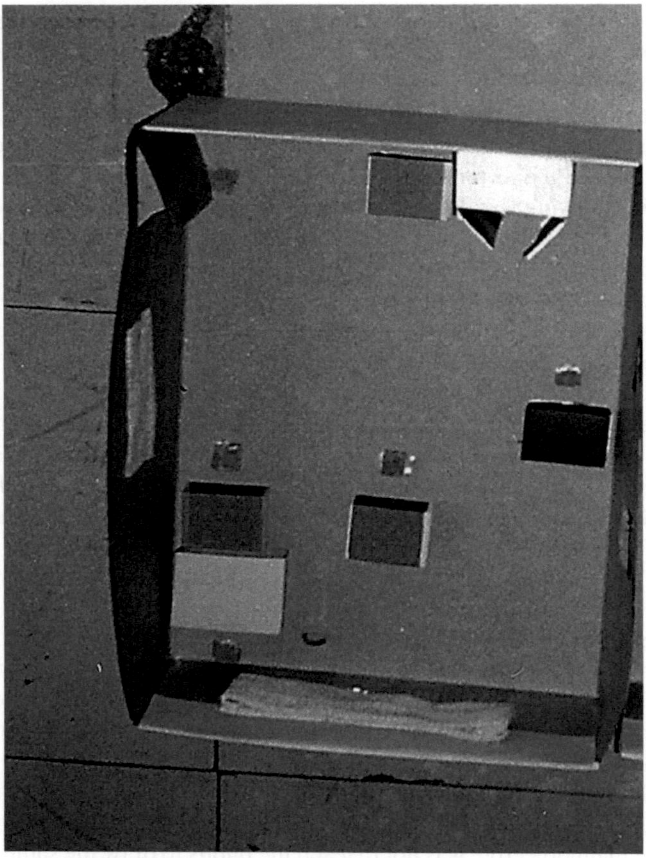

**Fig. 23.6** Laura's model of the classroom

Project activities took place, he used a crayon on paper and attempted to represent objects with which he had contact, such as the computers, floor and walls (Fig. 23.7).

After Bruno had explored a model representing the environment, he was asked to do a second representation of the same place. In the second drawing we see that the pupil seeks to represent the floor (number one in the drawing) with a graphic shape that resembles the real thing, but he cannot spatialize this shape on the paper and distribute the other items on it (Fig. 23.8).

Another activity that was proposed was the theme of how to draw *similarities between Brazil and Africa*. The drawing was free and the theme proposed was based on work already done with pupils in Regular Education and in the project. In his drawing the pupil ran the crayon over the paper again and explained that he had drawn *hunger, a drum/cylinder, a ball symbolizing the World Cup in Africa and industries*. The pupil demonstrates some knowledge of both territories, but again just renders a series of scribbled markings (Fig. 23.9).

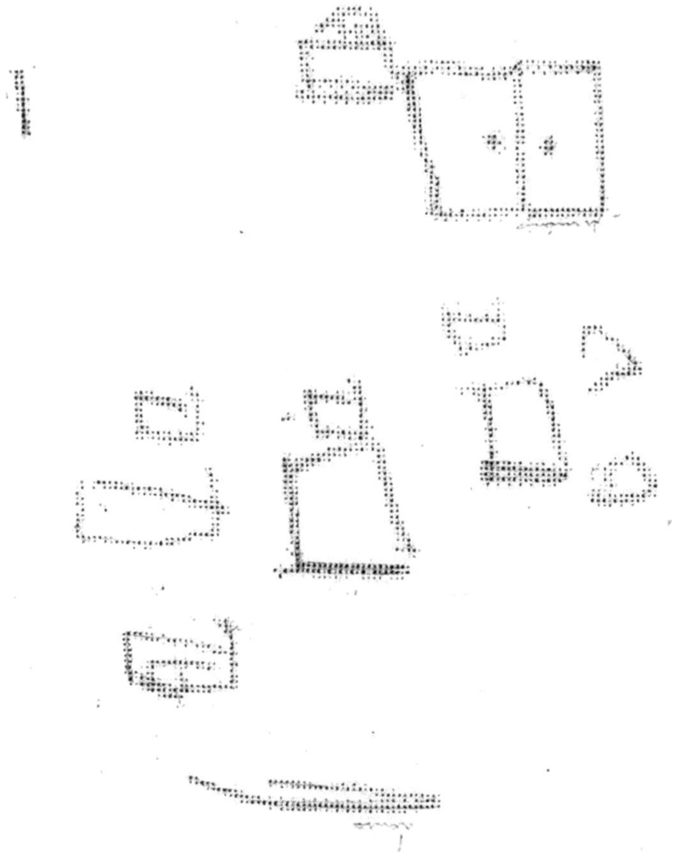

**Fig. 23.7** Laura's drawing of the classroom

In the study that we conducted regarding spatial organization with blind pupils at the Special School we had mixed results. In a representation done by Laura, there was not a representation of the shapes of the objects due to the need for successive integration of perceptions by means of touch, making it impossible to exploit them as a whole to understand their shapes. But we realized that in her drawing there were not just crayon scribbles in the attempt to represent the objects, as we found in the representations done by Bruno.

In the drawing produced by this pupil there is a spatial logic represented that fulfills the verbal logic: "The pupil visited the area weekly to attend Sunday Mass at the main Church, located on this square. After Mass, the pupil would go down the steps of the church, walked to the ice cream parlor, just across the street, near the Church" (Ventorini 2007) and had an ice cream accompanied by her parents (Fig. 23.11).

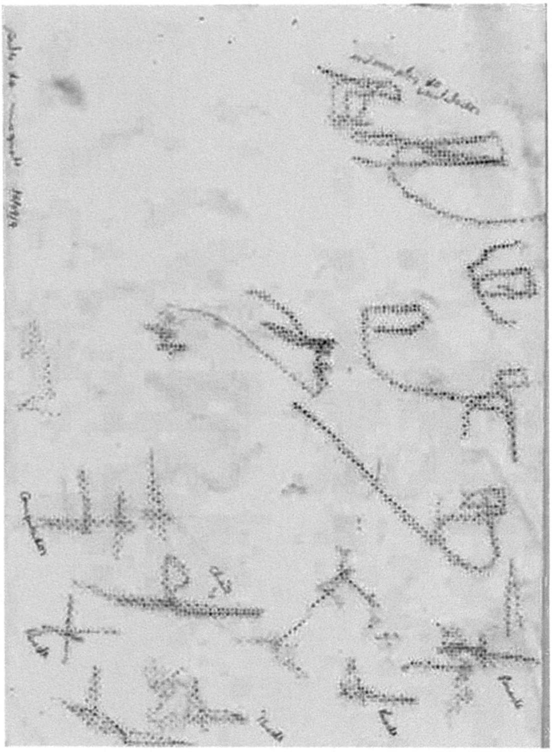

**Fig. 23.8** Bruno's mental map of the computer room

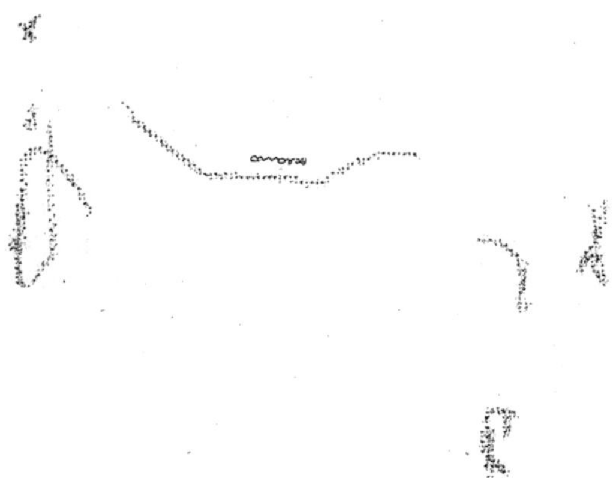

**Fig. 23.9** Drawing made after the manipulation of the model

The Church is represented by a square and a semi-triangle. In this representation the pupil uses concepts of contour lines for objects and figures with which she had experience, for example figures of a house, of churches etc. In her representation of

**Fig. 23.10** Drawing of the similarities between Brazil and Africa

the church the analysis indicates the use of concepts acquired in manipulating figures, since it was not possible to explore the object such as occurred with the dollhouse, which the pupil seeks to represent with the real shape of the object and not with the use of symbols.

The symbols are adopted for large objects, that cannot be explored by feeling them.

To represent the trees in the square she makes use of a simple figure of a tree and for the garden she uses the symbol of a flower. To represent the ice-cream parlor she uses two squares, which in fact symbolize the ice-cream parlor tables. Besides this, there are three ice creams, which represent her mother, her father and her. The symbols used to represent objects on the route are representations of her experience in this place in addition to the motor coordination skills and graphic memory acquired during the drawing activities.

We believe that this pupil acquired her drawing concept by having applied the knowledge acquired in her previous experience in this locality and learning how to

**Fig. 23.11** Laura's drawing of a route

draw to the challenge presented (representing a place in the town that she lived in). In her drawing the shape and size of objects were not represented because they are large and it was not possible to explore them by touch. The symbols used by Laura are not crayon scribbles on paper, but application of the graphic representations learned at the Special School.

The pupil does not know how many trees there are in the square, or what their height and shapes are, but represents them with a symbol of a tree. In order to represent the street and sidewalk she uses nearly circular shapes, indicating the spacing of her body. She represents tables at the ice cream parlor with those that she had contact, embracing the concepts for the development of squares. With regard to the steps she seeks to represent the differences in height, indicating the steepness of the place, and to achieve this she draws three almost elliptical shapes, one on top of the other.

This analysis coincides with the interpretation that we present about blind people's drawings in the introduction of this paper: "*The child's drawing is, thus, a system of representation. It isn't a copy of objects, but an interpretation of that which is real, done by the child in graphic language*" (Almeida 2010). Laura interprets reality and represents it by adopting the concepts of graphic language learned. It is not simple scribbling in crayon on paper, and nor they are copies of objects or figures.

The pupil at the Children's Rehabilitation Center does not have these memories, so he simply scribbles in crayon on the paper, without attributing a graphic shape to his mental image. Learning these memories is neither simple nor quick and does not just train motor skills, but it also gives sensory and psychic meaning to the mental act whose education and learning should be done in context.

In our research we concluded that the blind child develops this concept like any other child, acquiring the concept of permanent object, acquiring semantic memory and graphic act, assigning sensory and psychic meaning to the graphic act and demonstrating difficulties in drawing places or objects with which they have no significant experience.

## References

Almeida RD (2010) Do desenho ao mapa. Contexto, São Paulo

Batista CG (2005) Formação de conceitos em crianças cegas: questões teóricas e implicações educacionais. Psicologia: Teoria e Pesquisa 21(1):7–15, Brasília

Duarte MLB (2008) A imitação sensória-motora como possibilidade de aprendizagem do desenho por crianças cegas. Ciências e Cognição 13:14–26

Ochaíta E, Espinosa MA (2004) Desenvolvimento e intervenção educativa nas crianças cegas ou deficientes visuais. In: Marchesi A, Palácios J et al (eds) Desenvolvimento Psicológico e Educação: transtornos de desenvolvimento e necessidades educativas especiais, vol 3, 2nd edn. Artmed, São Paulo

Rubayo SC et al (2007) Discapacidad visual y destrezas manipulativas. ONCE, Madrid. http://www.once.es/serviciosSociales/index.cfm?navega=detalle&idobjeto=134&idtipo=1. Accessed 6 Oct 2010

Simón C, Ochaíta E, Huertas JA (1995) El sistema Braille: bases para su enseñanza-aprendizaje. Aprendizaje Comunicación, Lenguaje y Educación 28:91–102, Madrid

Ventorini SE (2007) A experiência como fator determinante na representação espacial do deficiente visual. Dissertation, IGCE, Universidade Estadual Paulista

Ventorini SE (2009) A experiência como fator determinante na representação espacial do deficiente visual. UNESP, São Paulo

# Chapter 24
# Tactile Cartography and Geography Teaching: LEMADI's Contributions

**Carla Cristina Reinaldo Gimenes de Sena and Waldirene Ribeiro do Carmo**

**Abstract** This paper presents an overview of the experiences on Tactile Cartography collected at LEMADI (Laboratory of Teaching and Didactic material of the Geography Department—Faculty of Philosophy, Letters and Human Sciences, University of São Paulo-USP). This is an attempt at displaying the contributions of this work to Geography teaching. LEMADI's researchers have been working for more than 20 years on producing, applying and evaluating geographic representations for teaching, orientation and mobility. The results from the first project as well as the materials already produced (maps, graphics, plans, illustrations, models and others) led to the formation of a permanent group attended by teachers, professionals on special education and visually disabled students. Experiences were shared by Brazilian and foreign researchers and institutions. LEMADI became a reference in Tactile Cartography not only for its tactile didactic materials used by students from elementary and high schools and college, teachers from public and private schools, parents, specialized teachers and blind people in general, but mainly for the methods developed by its researchers.

## 24.1 Introduction

In 1989, Professor Regina Araujo de Almeida (Vasconcellos) started a pioneering project at LEMADI in the area of tactile cartography in Brazil. Her research, "Tactile Map Production and the Visually Disabled: an evaluation of the stages of production and use of tactile maps" (1993) proposes a new method on Geography

C.C.R.G. de Sena (✉)
UNESP – Univ. Estadual Paulista, Campus de Ourinhos, Ourinhos, Brazil
e-mail: carla@ourinhos.unesp.br

W.R. do Carmo
Universidade de São Paulo, São Paulo, Brazil
e-mail: walcarmo@usp.br

L. Zentai and J.R. Nunez (eds.), *Maps for the Future*,
Lecture Notes in Geoinformation and Cartography 5,
DOI 10.1007/978-3-642-19522-8_24, 

teaching for the visually disabled, pointing out the role of graphic representations, mainly maps, in the process of space perception and in the acquisition of geographic notions.

In the period from 1990 to 1998, with the financial support of VITAE Foundation and from the University of São Paulo (Pró-Reitoria de Pesquisa), several projects were carried out under the coordination of Professor Regina Araujo Almeida at LEMADI. The group counted on the participation of geographers and students from the Geography course, including some who had support from a scholarship called "Bolsa-Trabalho" from the Social Assistance Coordination –COSEAS-USP.

The main objective of the research was to arouse the interest of visually disabled students in Geography and Cartography and, at the same time, to offer conditions for their assimilation of special concepts and information through touch, hearing and, occasionally, through residual vision. Thus, many didactic materials were made by associating colours with tactile language and using adapted writing for subnormal vision.

In the first project, Amazonia was used as a test area to analyze the methodology and to point out the problems. This work included the construction of a set of didactic materials, such as maps, models, plans, history books, an illustrated dictionary, activities and games, an illustrated time-line and teacher's manual, together with a text on Amazonia and its history since the sixteenth century.

The methodology was evaluated through tests with students from public and private elementary and high schools. This contact with visually disabled students made it possible to verify the efficacy of tactile graphic language in geography teaching, like the perception of space and the understanding of basic concepts such as proportion and scale, placement and orientation.

In the second project, the State of São Paulo was chosen for the application of the methodology already tested with Amazonia. At the same time, a training course was offered to help the practical use of our work. In the third project, an Atlas of the Continents was created.

The positive results from the first projects and the tactile didactic materials developed led to the formation of a permanent group to give support to teachers, professionals on special education and visually disabled students. Experiences started to be shared by Brazilian and foreign researchers and institutions.

## 24.2 International Projects

In 1994, a group of researchers from Chile, Argentina and Brazil started a project on the production, evaluation and application of tactile didactic materials. Firstly, a large bibliographic survey was carried out on the subject, which enabled the construction of a theoretical landmark. From that, it was possible to present proposals on the making of cartographic materials adapted to visually disabled children from elementary school and teachers from special education. With the financial support of institutions such as IPGH (Panamerican Institute of Geography and History) and the OEA (Organization of American States), several projects were carried out, and experiences will be presented by the international group.

The project "Diseño y Producción de Cartografia para las Personas Ciegas de America Latina" was developed in the period from 2002 to 2006 by researchers from Argentina, Brazil and Chile with the financial support of the OEA. It was coordinated by Professor Alejandra Coll from the Cartography Department of the Faculty of Humanities and Social Communication Technologies, Metropolitan University of Technology (UTEM), Santiago, Chile. Regarding the production of didactic materials and the organization of courses for teachers, the project counted on the collaboration of researchers from Cuyo National University of Argentina and from the LEMADI-DG-USP, Brazil in order to contribute to the improvement and diffusion of production and reproduction techniques of tactile graphic representations.

The main purpose of this project was to support visually disabled people in developing special abilities by the use of didactic and cartographic materials such as atlases, maps, charts, three-dimensional graphic systems and others in an attempt to improve their formal education and mobility.

This aim was achieved by the making and distributing products with thematic information (ecosystems, social, economic and cultural aspects) at several scales (global, continental, national, regional and local).

For the development of the project, an interdisciplinary group was formed composed by geographers, educators, sociologists, designers and others from the three participating countries. The whole group met at least once a year to exchange experiences, present results and define the next stages. The Metropolitan University of Technology (UTEM—Santiago, Chile) founded, with part of the project resources, the Latin American Tactile Cartography Center (CCAT). This Center became the place of coordination and developing the tactile graphic materials proposed (Figs. 24.1, 24.2, 24.3).

Throughout the 4 years (2002–2006), several tactile didactic cartographic materials were produced at several scales, followed by manuals with more than 150 tactile graphic representations:

- Set of physical, political, population density, vegetation and climatic maps. (Representations of the world, of Latin America and several member countries of the OEA).
- Didactic materials for geography teaching.
- Planet Earth general characteristics plan (water cycle, tectonic plates and others).
- Models for the teaching of geographic concepts.
- Handbook on geographic concepts, printed and in Braille, with the definition of each concept.
- Urban plans of Latin American capitals.
- Braille alphabet and printed maps.

One of the biggest challenges at the initial stage of the project was to define standards for the geographic representation. The discussions were based on

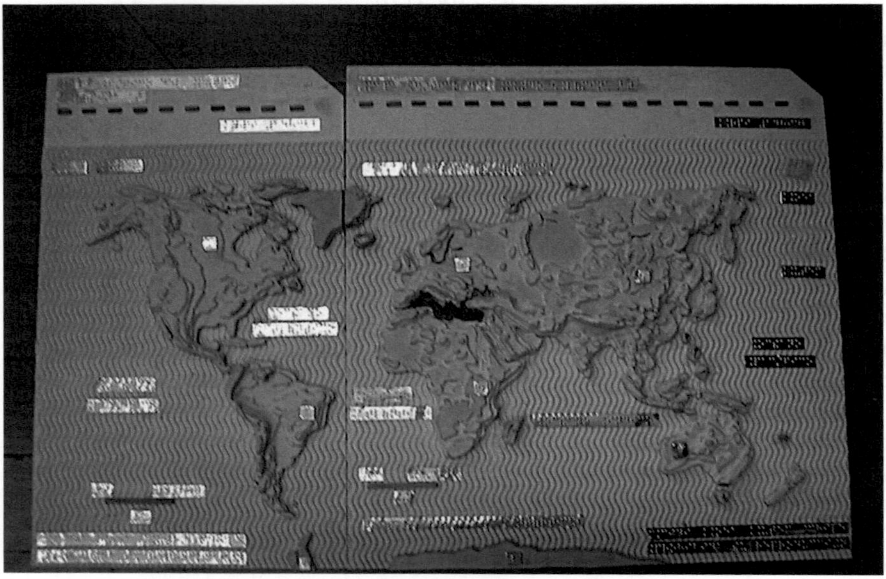

**Fig. 24.1** World physical map produced in the project

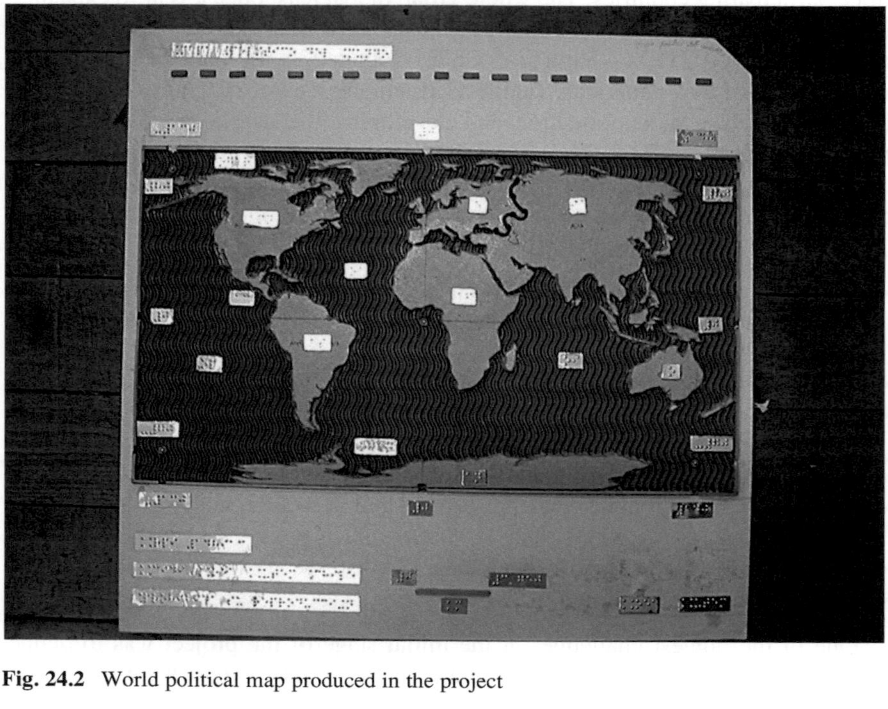

**Fig. 24.2** World political map produced in the project

**Fig. 24.3** Water cycle and continental drift produced in the project

professional experience and on the existing literature. Some of the patterns were limited by the parametres of thermophorm machine available at CCAT.

For this reason, it was decided that some standards would be applied for all maps of the project in order to make the reading and the evaluation of visually disabled students easier. These standards define the sort of tactile graphic representation produced at CCAT and distributed among the OEA members and project collaborators. It was established, for instance, that the North would be indicated by a dashed horizontal line placed on the top of the map. The maps showed, as frequently as possible, the North in this position, facilitating the students reading and comprehension of this variable or geographic notion.

As the convention for printed maps defines that the North (magnetic and geographic) is represented by an arrow. The Brazilian group proposed the keeping of the arrow for identifying the North direction. In this case, it would be represented with a vertical straight segment and an arrow in relief with the letter N in Braille, or a variation suggested by Vasconcellos (1993), in which the dashed line is complemented by an arrow.

However, after the evaluation of some maps by teachers specialized in visual disability, all the group researchers reinforced the idea of the use of the dashed line, confirming, then, the linear representation of the North as shown by Fig. 24.4.

It was decided that the scale would be graphically represented. The efficacy of the graphic scale in tactile maps has already been confirmed by Vasconcellos' research (1993). The concept of scale is abstract and its representation in a numerical form (proportion between real space and the represented one, using centimetres) makes it difficult to understand the idea of reduction of the map. This would demand another concept, the transformation of measures. However, the simplified scale representation may make the reading of the map and the comprehension of the concept of scale easy.

With the graphic scale, the user can explore the map with both hands: one of them fixed at the scale, and the other searching distances. In this way, one can establish a relation of proportion between the line segment of the scale and the represented distances.

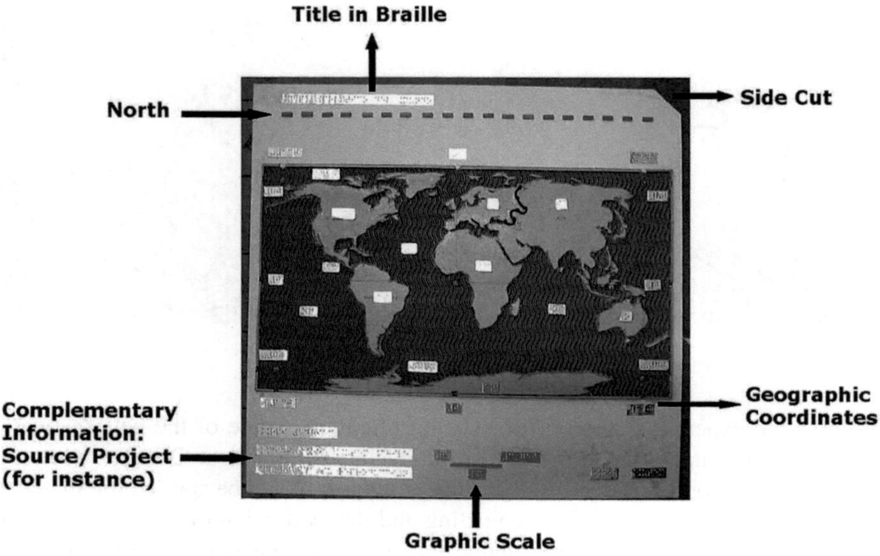


**Fig. 24.4** Tactile map with the proposed standardization

Besides, it was defined that all written information needed for reading the map would be made in Braille—title, scale values and the legend.

After the definition of the final originals for each participating country, data were collected on their population and climate for the compilation of graphics that complement the maps already produced in order to help the teaching of these subjects (Figs. 24.5, 24.6).

The last stage focused on the search of cartographic bases for producing at least one urban (downtown) map or a map of the most relevant area of the cities from 18 Latin American countries. These maps were made for orientation and moving in the city. This is why the maps were produced at large scale, which helped identifying the important buildings. Every country chose a central area of the capital, except for Brazil, which represented a part of São Paulo downtown.

The maps were made in collage and copied on transparent plastic in order to add the printed information to the whole. Beyond the urban maps, models of classrooms were made by the schools that collaborated to the evaluations of the first stage. The result was 18 urban maps, 4 enlargements of these maps (La Paz, São Paulo, Quito and Lima) and 16 maps from schools that participated in the classroom project.

For constructing the tactile graphic representations, tests on the durability and heat resistance of the materials were made. The thermophorm machine, which reproduces the maps, works with a vacuum and heat system, therefore, the originals must be resistant. Thus, a large number of copies can be made from the originals. After the tests, several types of card paper, metals, wood sticks, sandpaper and several threads were selected for representing information in a tactile shape. Braille

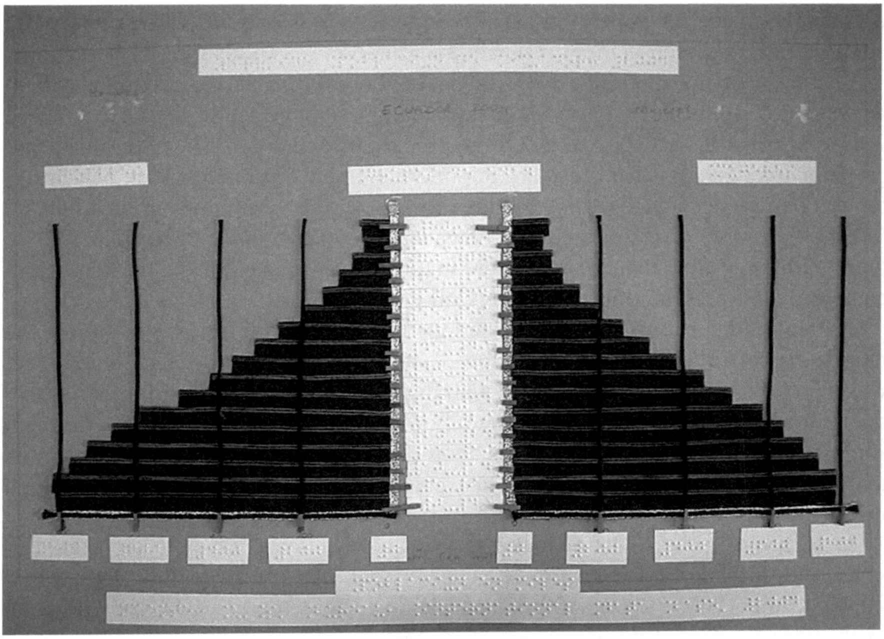

Fig. 24.5 Representation of the equatorial pyramid—collage

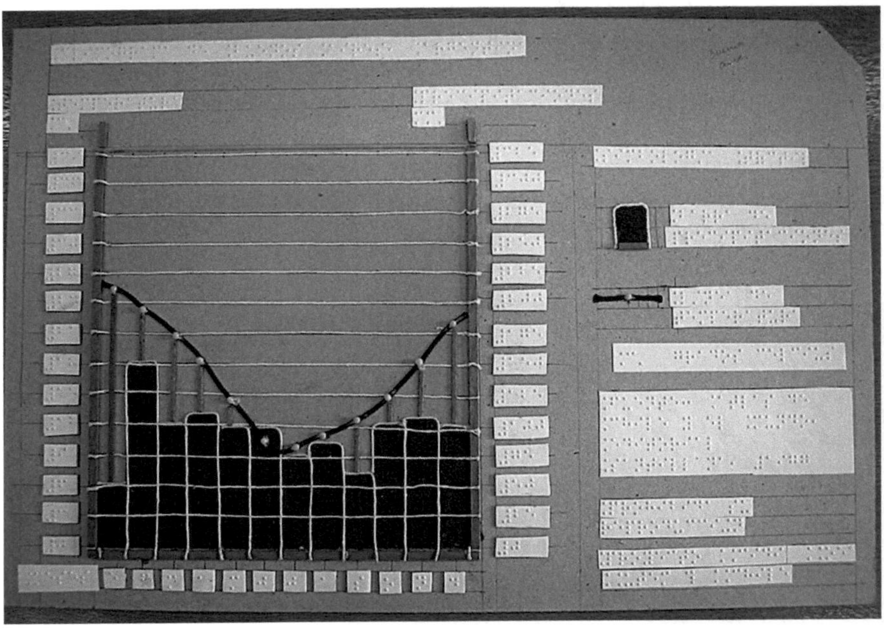

Fig. 24.6 Climatic diagram of Buenos Aires—collage

was firstly made on paper, but as its durability was low, it was replaced by aluminium foil, which is more resistant and durable.

The tests of the materials were made in schools that have visually disabled students (16 from the 18 Latin American countries belonging to the OEA). These tests confirmed the efficacy of tactile graphic representations, especially maps and graphics, recognizing that these are important elements in the teaching of Cartography, Geography and History, because they help the communication of special information. 85% of the participants considered the importance of the materials to their learning, stating that it is a pleasure to touch, easy to handle, and the maps have clear and well-presented information (Coll e Pino 2007).

A great number of students suggested the insertion of printed information in the maps in relief in order to make the low-vision people's reading easy. This aspect was also pointed out by the teachers who evaluated the material, emphasizing that maps in relief and paint help the integration of pupils, since the visually disabled student uses the same material as his "non-disabled" mate. It must be highlighted that the urban maps made in the last stage of the project considered printed information and were reproduced on transparent PVC.

In addition to the material making, five qualification courses were offered to several Latin American teachers. The courses took place in Argentina (December/2004), Chile (2003, 2005, 2006) and Mexico (September/2004), providing an exchange of information and experiences that was fundamental to the definition of methodologies for the application of the materials in classrooms. More than 50 professionals were trained throughout the project. All the information acquired in the courses was taken by these teachers to their mother countries.

As not all teachers who are specialized in visual disability are necessarily experts in Geography, it was necessary to introduce basic notions on cartography to deal with the techniques of construction and methodologies of the use of the materials.

The courses were organized in this manner: cartography basic notions (scale, geographic coordinates, and graphic symbology); techniques for the construction of tactile maps (collage and aluminium), creation of games in EVA, and tactile maps reading methodology (Fig. 24.7).

**Fig. 24.7** Teachers producing materials during the course

The participants in the courses gave a very important contribution to the development of the research on Tactile Cartography. They pointed out the difficulties of finding tactile graphic representations in their countries of identifying the students' necessities such as their need of specific graphic representations, the representation of a specific neighbourhood where services considered important by them are placed (school, hospital, therapist, and others). Also, the detailing of a political map of their countries is important for the comprehension of their own country. That reinforces the importance of the teachers' qualification for the production and using of tactile didactic materials.

Considering that a great part of the schools that visually disabled students attend have access to few resources for the materials acquisition, the use of simple and low-cost materials was emphasized. The alternative use of different papers, recycled tissues and rests in general made it possible to improve the representations used in classrooms by some of the participating teachers.

The second project to be remarked is called "*Integrando Los Sentidos en el Manejo de la Información Geoespacial, Mediante la Cartografía Táctil, con Especial Énfasis en las Personas Ciegas y Sordas de América Latina*". It was developed in the period from 2007 to 2009 with the financial support of IPGH, in a partnership with the University of São Paulo (LEMADI)—Brazil, the Metropolitan University of Technology (UTEM), Latin American Tactile Cartography Center (CCAT)—Santiago, Chile, and the School Nuestra Señora del Carmen—Cusco, Peru.

The general purpose of this project was the adaption of three-dimensional cartographic material and the developing of new processes for the construction of geographic information models to people with hearing and/or visual disability from Latin America. The project was followed by an intensive discussion and the set up of a theoretical landmark on Geography teaching for visually and/or hearing disabled people by the use of Tactile Cartography.

Physical and political maps of the metropolitan regions of Santiago, Chile, Cusco and São Paulo were made by the use of the collage technique. The bases were reproduced on transparent plastic and the representations were evaluated afterwards by visually disabled students from the three countries.

It is important to point out that the standardization discussed and evaluated in the previous projects was kept and the new evaluations reinforced the efficacy of the use of patterns to the tactile graphic representations. Besides, this project reinforced the necessity of the creation of resources that can be used simultaneously by visually disabled people and other sorts of disabled people.

The project attempted at making the representations on transparent plastic, which superposes a printed copy with written information beyond the graphic ones (Figs. 24.8, 24.9).

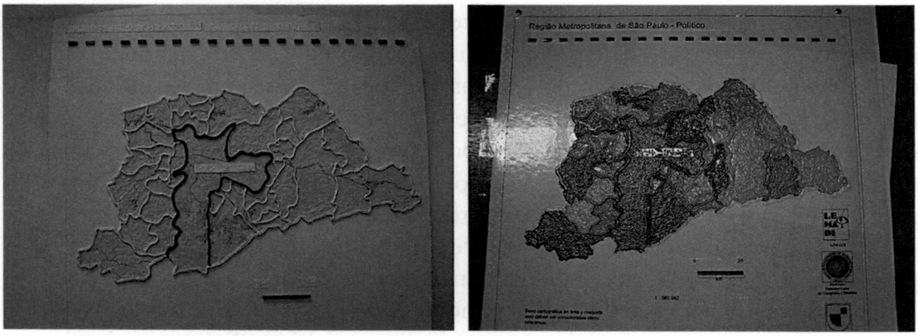

**Fig. 24.8** Metropolitan region of São Paulo—political—produced during the project (original and copy in thermoform)

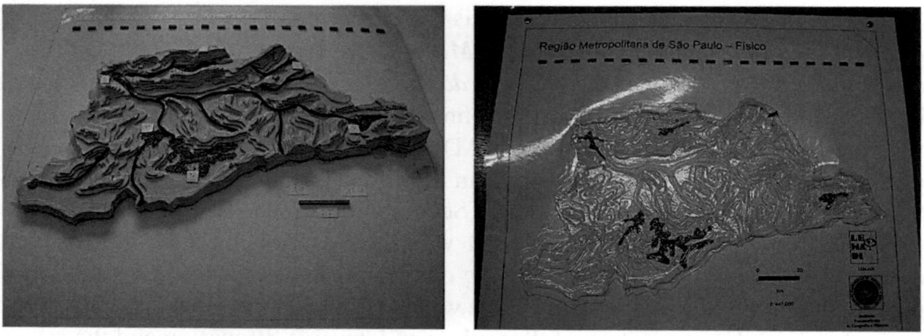


**Fig. 24.9** Metropolitan region of São Paulo—physical—produced during the project (original and copy in thermoform)

## 24.3 Final Considerations

The experience acquired in the area of cartography for visually disabled people and a positive evaluation of the tactile resources built until now, also in schools for auditive disabled people, demonstrated that the tactile didactic materials can and must be used by people with hearing and/or visual disability in the learning of Geography, motivating the learning through senses: touch, sight and hearing.

The materials produced make it possible to work in an integrated manner, with children and young visually disabled people, and with "no-disabled" students, starting a new area for the cartographic discipline development.

Beyond school, the media frequently uses different types of graphic representations (maps and mainly graphics) in newspapers, magazines, Internet and on TV to illustrate or explain several subjects. The visually disabled ones receive, thus, this sort of information through an oral representation, which is not always convenient. The Tactile Cartography may reduce and overcome this information restriction.

The use of graphic resources in relief, therefore, provides the overcoming of the informational barriers, contributes to the inclusion of the disabled ones in school, at work and in daily-life. This way, the tactile cartography benefits those who depend on the touch and on the hearing to apprehend images, to use maps and to comprehend graphics. Besides, it can be used in every classroom, in every school, with all students.

Developing materials and methods to help Geography teaching for all, independently of its differences, by respecting each one's necessities, is an important step for the concept of inclusion.

This inclusion can only be put into practice with the partnership between researchers and teachers. In this sense, the project experiences provide the organization of courses and workshops for the diffusion of techniques and methodologies on the using of didactically adapted materials, by the constant exchange of experiences on the theme and the discussion of its process at school.

LEMADI-DG-USP is nowadays an important space in the area of Tactile Cartography, a space to the development of discussions on inclusion, where the whole community can benefit on the materials and on the results reached up to this moment.

## References


Almeida R (2007) A cartografia tátil no ensino de geografia: teoria e prática. In: Almeida RD (ed) Cartografia escolar. Contexto, São Paulo

Carmo WR (2010) Cartografia tátil escolar: experiências com a construção de materiais didáticos e com a formação continuada de professores. Dissertação de Mestrado. Departamento de Geografia, FFLCH, Universidade de São Paulo

Coll A (2004) Contribución al proceso de enseñanza aprendizaje en la educación de personas ciegas por medio de la cartografía. In: Informe final año 2004 proyecto diseño y producción de cartografía para las personas ciegas de América Latina. Anexo IV publicaciones

Coll A, Pino F (2007) Impacto de la cartografia táctil em la enseñanza de la geografia em America Latina. In: XXIII International Cartographic Conference, Moscow, 4–10 August 2007. ICA

Edman PK (1992) Tactile graphics. American Foundation for the Blind, New York

Sena CCRG de (2002) O estudo do meio como instrumento de ensino de geografia: desvendando o pico do jaraguá para deficientes visuais. Dissertação de Mestrado. Departamento de Geografia, FFLCH, Universidade de São Paulo

Sena CCRG de (2008) Cartografia tátil no ensino de geografia: uma proposta metodológica de desenvolvimento e associação de recursos didáticos adaptados a pessoas com deficiência visual. Tese de Doutorado. Departamento de Geografia, FFLCH—USP

Sena CCRG de, Carmo WR (2005a) Tactile map production for the visually impaired user: experiences in Latin America. In: XXII International Cartographic Conference, A Coruña, 9–16 July 2005. ICA

Sena CCRG de, Carmo WR (2005b) Uso de maquetes no ensino de conceitos de Geografia Física para deficientes visuais. In: XXVI congreso nacional y XI internacional de geografia, Santiago de Chile

Sena CCRG de, Carmo WR (2005c) Uso de maquetes no ensino de conceitos de geografia física para deficientes visuais. In: XI simpósio brasileiro de geografia física aplicada, São Paulo

Sena CCRGde, Carmo WR (2005d) Produção de mapas para portadores de deficiência visual da América Latina. In: X encontro de geógrafos da América Latina, São Paulo
Vasconcellos RA (1993) Cartografia tátil e o deficiente visual: uma avaliação das etapas de produção e uso do mapa. Tese de Doutorado. Departamento de Geografia. FFLCH-USP. São Paulo